RESTRICTED   AN 01-45HB-1

Pilot's Handbook
of
Flight Operating Instructions

NAVY MODEL

F4U-4 Airplane

**THIS PUBLICATION SUPERSEDES AN 01-45HB-1 DATED 1 OCTOBER 1944
REVISED 15 DECEMBER 1944**

Appendix I of this publication shall not be carried on aircraft on combat missions or where there is a reasonable chance of its falling into the hands of the enemy

NOTICE.—This document contains information affecting the national defense of the United States within the meaning of the Espionage Act, 50 U. S. C., 31 and 32, as amended. Its transmission or the revelation of its contents in any manner to an unauthorized person is prohibited by law.

RESTRICTED AN 01-45HB-1

Pilot's Handbook of Flight Operating Instructions

NAVY MODEL
F4U-4 Airplane

THIS PUBLICATION SUPERCEDES AN 01-45HB-1 DATED 1 OCTOBER 1944
REVISED 15 DECEMBER 1944

Appendix I of this publication shall not be carried on aircraft on combat missions or where there is a reasonable chance of its falling into the hands of the enemy

NOTICE.—This document contains information affecting the national defense of the United States within the meaning of the Espionage Act, 50 U. S. C., 31 and 32, as amended. Its transmission or the revelation of its contents in any manner to an unauthorized person is prohibited by law.

1 April 1945

RESTRICTED

Published under joint authority of the Commanding General, Army Air Forces, the Chief of the Bureau of Aeronautics, and the Air Council of the United Kingdom

THIS PUBLICATION MAY BE USED BY PERSONNEL RENDERING SERVICE TO THE UNITED STATES OR ITS ALLIES

Navy Regulations, Article 76, contains the following statements relating to the handling of restricted matter:

Par. (9) (a). Restricted matter may be disclosed to persons of the Military or Naval Establishments in accordance with special instructions issued by the originator or other competent authority, or in the absence of special instructions, as determined by the local administrative head charged with custody of the subject matter.

(b) Restricted matter may be disclosed to persons of discretion in the Government service when it appears to be in the public interest.

(c) Restricted matter may be disclosed, under special circumstances, to persons not in the Government service when it appears to be in the public interest.

The Bureau of Aeronautics Aviation Circular Letter No. 31-44 contains the following paragraph relative to the use of technical aeronautics publications:

Par. 8. *Distribution to All Interested Personnel.*—In connection with the distribution of aeronautical publications within any activity, it should be borne in mind by the officers responsible for such distribution that technical publications are issued specifically for use not only by officer personnel, but more particularly by responsible civilian and enlisted personnel working with or servicing equipment to which the information applies.

Paragraph 5 (d) of Army Regulation 380-5 relative to the handling of restricted printed matter is quoted below:

(d) *Dissemination of restricted matter.*—The information contained in restricted documents and the essential characteristics of restricted material may be given to any person known to be in the service of the United States and to persons of undoubted loyalty and discretion who are cooperating in Government work, but will not be communicated to the public or to the press except by authorized military public relations agencies.

These instructions permit the issue of restricted publications to civilian contract and other accredited schools engaged in training personnel for Government work, to civilian concerns contracting for overhaul and repair of aircraft or aircraft accessories and to similar commercial organizations.

LIST OF REVISED PAGES ISSUED

NOTE.—A heavy black vertical line to the left of the text on revised pages indicates the extent of the revision. This line is omitted where more than 50 percent of the page is involved.

Vought F4U-4 Corsair
Pilot's Flight
Operating Instructions
©2006, 2009 Periscope Film LLC
All Rights Reserved

www.PeriscopeFilm.com
ISBN 978-1-4116-8960-2

ADDITIONAL COPIES OF THIS PUBLICATION MAY BE OBTAINED AS FOLLOWS:

AAF ACTIVITIES.—Submit requisitions to the Commanding General, Fairfield Air Service Command, Patterson Field, Fairfield, Ohio, Attention: Publications Distribution Branch, in accordance with AAF Regulation No. 5-9. Also, for details of Technical Order distribution, see T. O. No. 00-25-3.

NAVY ACTIVITIES.—Submit requests to Chief, BuAer, Navy Department, Washington, D. C., Attn.: Publications Section on order form NAVAER-140. For complete listing of available material and details of distribution see Naval Aeronautic Publications Index, NavAer 00-500.

RESTRICTED

RESTRICTED
AN-01-45HB-1

INDEX

Pilots Handbook

PRINCIPAL SECTIONS

FLIGHT OPERATING INSTRUCTIONS

for

U.S. NAVY MODEL

F4U-4
CORSAIR

Principal Sections

1	DESCRIPTION	*Page 1*
2	NORMAL OPERATING INSTRUCTIONS	*Page 23*
3	OPERATING DATA	*Page 39*
4	EMERGENCY OPERATING INSTRUCTIONS	*Page 41*
5	OPERATIONAL EQUIPMENT	*Page 45*
	OPERATING CHARTS, TABLES, CURVES AND DIAGRAMS	*Page 53*

RESTRICTED

TABLE OF CONTENTS

	Page
List of Illustrations	iii
List of Charts, Tables and Curves	iii

Section I
DESCRIPTION

1. Airplane 1
   a. General 1
   b. Engine 1
   c. Propeller 1
2. Controls Description 6
   a. General 6
   b. Power Plant Controls 6
   c. Fuel System Controls 8
   d. Oil System Controls 10
   e. Hydraulic System Controls 13
   f. Trim Tab Controls 20
   g. Balance Tabs 21
   h. Miscellaneous Controls and Equipment 21

Section II
NORMAL OPERATING INSTRUCTIONS

1. Before Entering the Cockpit 23
2. Entrance to Closed Airplane 23
3. On Entering the Cockpit 24
4. Fuel System Management 24
5. Oil System Management 26
6. Starting Engine 27
7. Warm-up and Ground Test 28
8. Scramble Take-off 29
9. Taxiing Instructions 30
10. Take-off 30
11. Engine Failure During Take-off 31
12. After Take-off 31
13. Climb and Level Flight 31
14. General Flying Characteristics 31
15. Maneuvers 34
16. Stalls 35
17. Spins 35
18. Permissible Acrobatics 36
19. Diving 36
20. Approach and Landing 36
21. Stopping of Engine 37
22. Before Leaving Cockpit 38
23. Mooring 38

Section III
OPERATING DATA

1. Airspeed Correction Table 39

Section IV
EMERGENCY OPERATING INSTRUCTIONS

1. Emergency Egress 41
2. Emergency Landing Gear Operation 41
3. Emergency Wing Flap Operation 42
4. Life Raft 43
5. Engine Failure During Flight 43
6. Forced Landings 43
7. Electrical Fire 44
8. Generator System Failure 44

Section V
OPERATIONAL EQUIPMENT

1. Operation of Oxygen Equipment 45
2. Operation of Radio, Communication and Navigation Equipment 46
3. Operation of Electrical Equipment 48
4. Operation of Armament 50

Appendix I

Operating Charts, Tables, Curves and Diagrams 53

LIST OF ILLUSTRATIONS

Fig. No.	Section I	Page
1.	The Corsair	iv
2.	Cockpit—Forward	2
3.	Cockpit—Left Hand Side	3
4.	Cockpit—Right Hand Side	4
5.	Main Instrument Panel	5
6.	Center Control Panel	5
7.	Engine Control Unit	6
8.	Fuel System Control Diagram	9
9.	Oil System Control Diagram	11
9A.	Oil System Warm-up Circuit	12
9B.	Oil Dilution System	12
10.	Inclined Panel—Left Hand Shelf	13
11.	Wing Lock Pin Indicator System	15
12.	Hydraulic System—Schematic	16
13.	Landing Gear Hydraulic System Control Diagram	17
14.	Dive Brake Hydraulic System Control Diagram	17
15.	Arresting Hook Hydraulic System Control Diagram	18
16.	Wing Flap Hydraulic System Control Diagram	18
17.	Wing Folding Hydraulic System Control Diagram	19

Fig. No.		Page
18.	Cooling Flaps Hydraulic System Control Diagram	19
19.	Gun Charging Hydraulic System Control Diagram	20
20.	Trim Tab Controls	20
21.	Cabin Control	21
22.	Tail Wheel Lock and Manual Drop Tank Release	21
	Section II	
23.	Fuel Tank Selector	24
	Section IV	
24.	Cabin Emergency Release	41
25.	Emergency Landing Gear Extension System	42
	Section V	
26.	Oxygen System	45
27.	Radio and Communications Controls	47
28.	Electrical Controls	49
29.	Gun Switch Box	51
30.	Rocket and Bomb Switch Box	51
	Appendix I	
31.	Protection Against Gunfire	53

LIST OF CHARTS, TABLES AND CURVES

	Page
Chart of Comparisons	1
Airspeed Limitations	23
Operating Flight Strength Diagram	33
Stalling Speeds	35
Air Speed Correction Table	39
Power Plant Chart	40
Oxygen Consumption Table	46

	Page
Take-off, Climb and Landing Chart	54
Flight Operation Instruction Chart—1	55
Flight Operation Instruction Chart—2	56
Flight Operation Instruction Chart—3	57
Angle of Attack vs. Dive Angle Curves	58
Angle of Attack vs. Air Speed Curves	59
Engine Calibration Curve	61

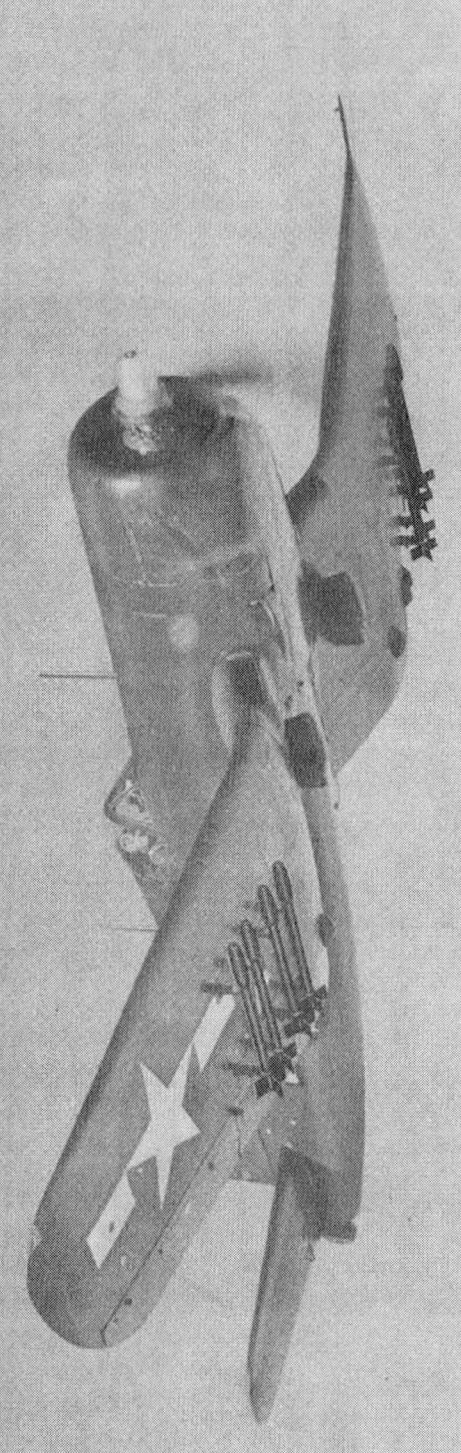

Figure 1 — The Corsair

RESTRICTED
AN-01-45HB-1

Section 1
Paragraph 1

Section I
DESCRIPTION

MAIN DIFFERENCES — F4U-1, F3A-1, FG-1 AND F4U-4 AIRPLANES						
MODEL DESIGNATION	EXTERNAL RECOGNITION FEATURES	MAXIMUM ARMAMENT*	MAXIMUM FUEL CARRYING CAPACITY	NORMAL GROSS WEIGHT	COCKPIT	ENGINE SERIAL NUMBER
F4U-1 F3A-1 FG-1	3-BLADED PROPELLER CIRCULAR ENGINE COWL	6—.50 CAL. MACH. GUNS 2400 ROUNDS OF AMM. 8—5-INCH ROCKETS 1—1000 LB. BOMB	MAIN FUEL TANK......237 GALS. TWO WING TANKS..126 GALS. ONE DROP TANK....170 GALS.	1200 LBS.		R-2800-8 OR R-2800-8W
F4U-1D FG-1D	3-BLADED PROPELLER CIRCULAR ENGINE COWL TWIN PYLONS ON WING CENTER SECTION	6—.50 CAL. MACH. GUNS 2400 ROUNDS OF AMM. 8—5-INCH ROCKETS 2—1600 LB. BOMBS	MAIN FUEL TANK......237 GALS. THREE DROP TANKS..510 GALS.	1200 LBS.		R-2800-8W
F4U-1C	3-BLADED PROPELLER CIRCULAR ENGINE COWL TWIN PYLONS ON WING CENTER SECTION	4—20 MM. CANNONS 924 ROUNDS OF AMM. 8—5-INCH ROCKETS 2—1600 LB. BOMBS	MAIN FUEL TANK......237 GALS. THREE DROP TANKS..510 GALS.	1200 LBS.		R-2800-8W
F4U-4	4-BLADED PROPELLER NON-CIRCULAR ENGINE COWL WITH AIR DUCT IN LOWER LIP TWIN PYLONS ON WING CENTER SECTION	6—.50 CAL. MACH. GUNS 2400 ROUNDS OF AMM. 8—5-INCH ROCKETS 2—1600 LB. BOMBS	MAIN FUEL TANK......233 GALS. TWO DROP TANKS....340 GALS.	12500 LBS.	RE-ARRANGED	R-2800-18W

*NOT NECESSARILY CARRIED SIMULTANEOUSLY.

1. AIRPLANE.

a. GENERAL.—The Model F4U-4 Airplane is manufactured by Chance Vought Aircraft Corporation, Division of United Aircraft Corporation, Stratford, Connecticut. The airplane is a single-engine, single-seat, low-wing monoplane designed as a carrier and land based fighter, and is equipped for operation as a long range fighter when carrying a droppable fuel tank, or as a fighter-bomber when carrying either one or two bombs. Provision is made for carrying six .50-caliber Browning machine guns and eight rockets. The approximate gross weight is 12,450 pounds with full ammunition loading and full main fuel tank, but with no external loading. The approximate overall dimensions are as follows:

Length ....33 ft.
Height ....13 ft.
Span ....41 ft.
Height with wings folded............16 ft.

b. ENGINE. — Pratt and Whitney Double Wasp; R-2800-18W; two stage supercharged; two speed auxiliary stage; propeller reduction gearing is .45:1. The engine is equipped with an automatic spark advance. See paragraphs 2.b.(4), this section, and 7.c.(10), Section II.

c. PROPELLER.—Hamilton Standard Hydromatic; four aluminum alloy blades (6501A-0); hub (24E60); thirteen feet, two inches diameter.

RESTRICTED

1

1. Main instrument panel
2. Gun switch box
3. Gun sight
4. Rocket and bomb switch box
5. Cabin control
6. Control stick
7. Electric and radio control box
8. Main tank pressure release
9. Center control panel
10. Rudder pedals
11. Brake pedals

Figure 2 — Cockpit — Forward

RESTRICTED
AN-01-45HB-1

Section 1

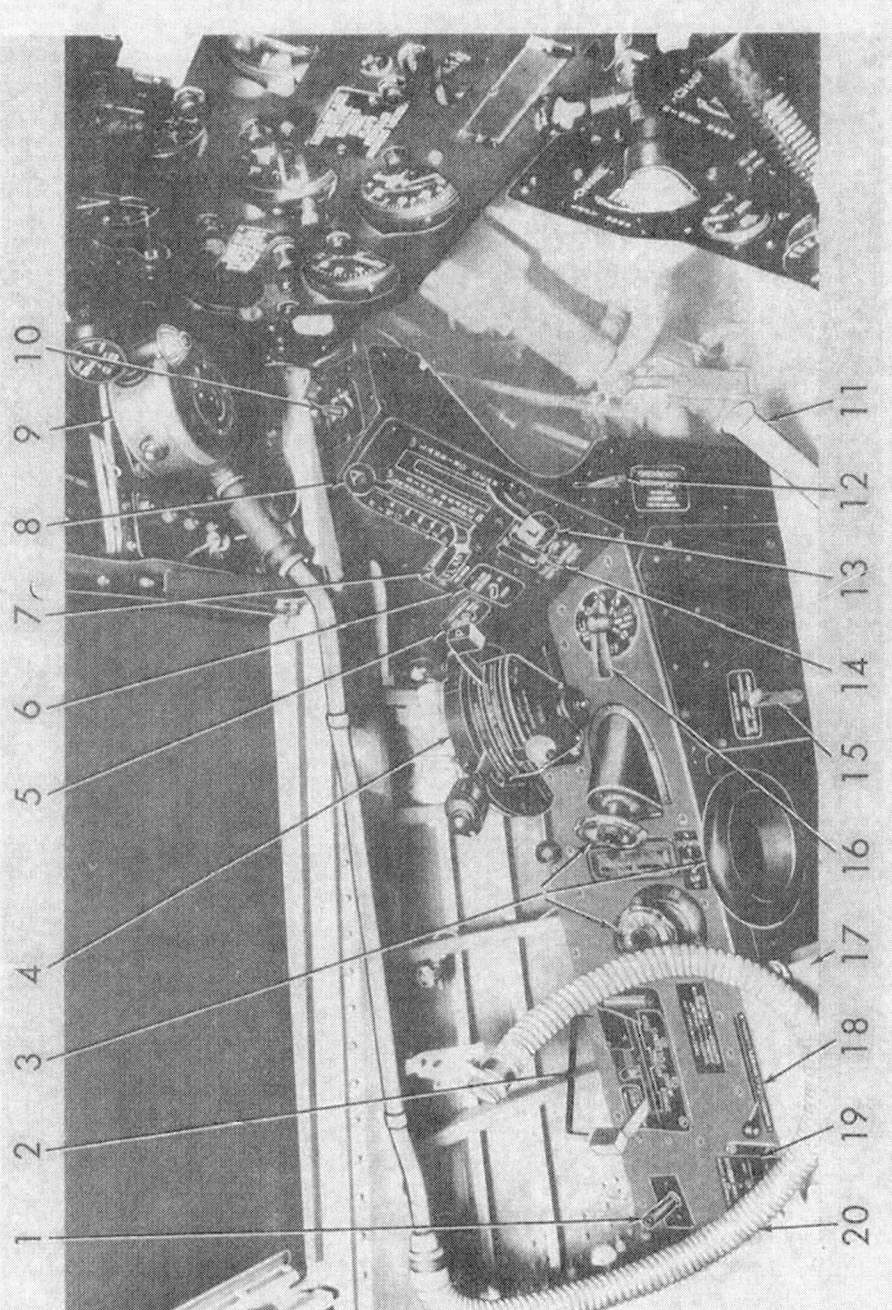

1. Tow target release
2. Wing folding and locking controls
3. Trim tab controls
4. Engine control unit
5. Fuel transfer switch
6. Booster fuel pump switch
7. Wing flap control
8. Arresting hook control
9. Diluter-demand oxygen regulator
10. Cowl flap switch
11. Hydraulic hand pump
12. Landing gear lock override control
13. Landing gear and dive brake control
14. Landing gear indicators
15. Hand pump selector
16. Fuel selector
17. Map case
18. Tail wheel lock control
19. Manual bomb and drop tank release controls
20. Oxygen tube

Figure 3 — Cockpit — Left Hand Side

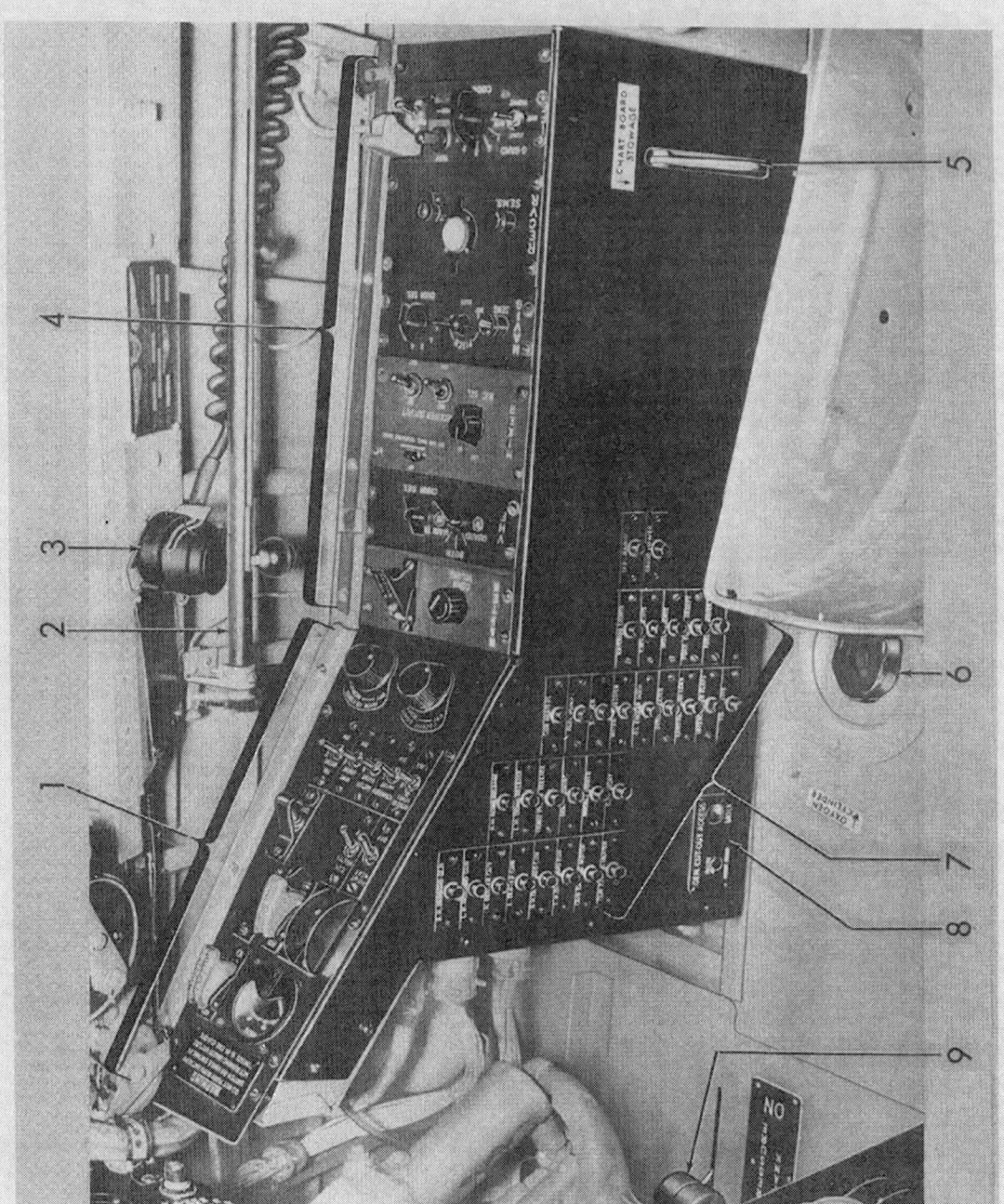

1. Electrical controls (See figure 28)
2. Retractable homing antenna control
3. Hand microphone
4. Radio and communications controls (See figure 27)
5. Chart board stowage
6. Oxygen bottle
7. Electrical system circuit breakers
8. Generator cutout access hole
9. Main tank manual pressure release

Figure 4 — Cockpit — Right Hand Side

Figure 5 — Main Instrument Panel

1. Main instrument board
2. Cover assembly
3. Compass indicator
4. Air speed indicator
5. Windshield defroster control
6. Gyro horizon
7. Climb indicator
8. Elapsed time clock
9. Fuel quantity gage
10. Cylinder temperature gage
11. Engine gage unit
12. Turn and bank indicator
13. Directional gyro
14. Altimeter
15. Manifold pressure gage
16. Tachometer
17. Ignition switch

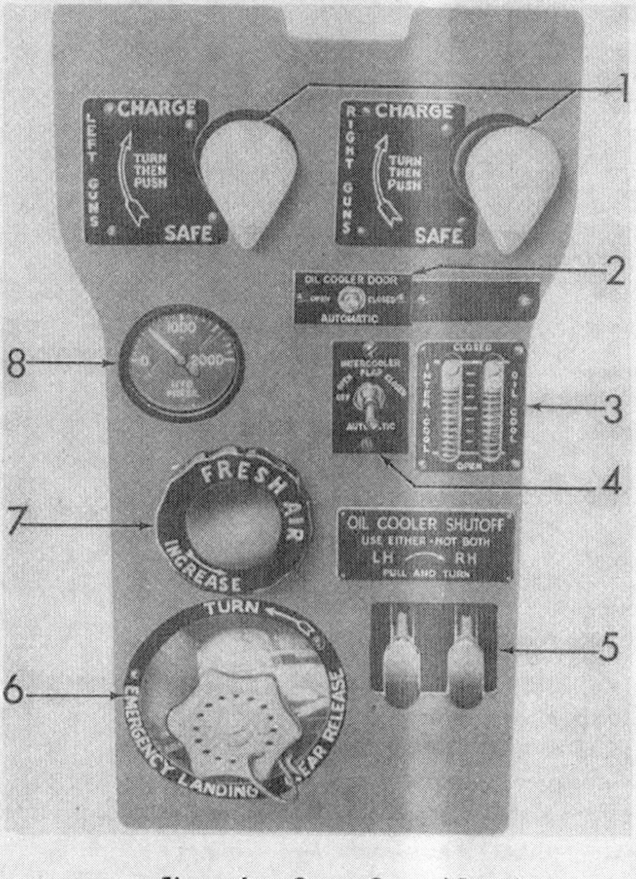

1. Gun charging controls
2. Oil cooler door control
3. Oil cooler and intercooler indicators
4. Intercooler flap control
5. Oil cooler shut-off controls
6. Landing gear CO_2 emergency release
7. Fresh air control
8. Hydraulic pressure gage

Figure 6 — Center Control Panel

2. CONTROLS DESCRIPTION

a. GENERAL.—In addition to the conventional surface controls, the cockpit controls consist of power plant, fuel system, oil system, hydraulic system and other miscellaneous controls. The location and operation of each control is described in the following paragraphs.

b. POWER PLANT CONTROLS.—The throttle, mixture, propeller governor, and supercharger controls are arranged in a unit installed on the left side of the cockpit as shown in figures 3 and 7. Each control moves through a quadrant in operation. The engine control unit is marked plainly with the name and correct positioning of all control levers mounted thereon. A friction adjustment knob is provided on the inboard side of the engine control unit. When the knob is tightened, movement of the engine controls will be stiffer, thus making it possible to prevent the controls from creeping. Other controls necessary to proper engine operation are the cowl flap, intercooler flap and oil cooler flap controls (see figure 18) and the starter and ignition switches.

Figure 7 — Engine Control Unit

(1) THROTTLE CONTROL. The throttle control is located on the engine control unit on the left hand side of the cockpit.

(2) WATER INJECTION LIMIT SWITCH.—In addition to its normal function, the throttle control operates the water injection limit switch and turns on the water injection equipment. Refer to Section II, paragraphs 4.*i.* and 14.*d.* for information concerning the water injection system and its operation.

(3) MIXTURE CONTROL. — The R-2800-18W engine is equipped with a Bendix-Stromberg PR-58E2 down-draft injection carburetor with automatic mixture control. The mixture control has three positions, "AUTO RICH," "AUTO LEAN," and "IDLE CUT-OFF," these positions being marked plainly on the side plate of the engine control unit. Fuel will be discharged from the carburetor at any fuel pressure above five pounds per square inch when the mixture control is not in the "IDLE CUT-OFF" position. To prevent flooding of the engine through inadvertent use of the booster fuel pump, the mixture control shall always be left in "IDLE CUT-OFF" whenever the engine is not running.

Note

When adjusting the mixture control, make sure that the control is set properly by feeling for the "notch" in the carburetor which indicates correct positioning of the carburetor lever. This, however, applies only to the "AUTO LEAN" position since "AUTO RICH" and "IDLE CUT-OFF" are the extreme positions of the mixture control lever. A notch is provided in the engine control unit for the "AUTO RICH" position of the mixture control handle.

(4) AUTOMATIC SPARK ADVANCE.

(a) The engine is equipped with a two-position automatic spark advance for providing better fuel economy at cruising powers. At idle, the spark is in the 20 degree (normal spark advance) position. When the engine speed reaches the range between 800 and 1500 rpm, the spark is advanced automatically to the 35 degree (cruising spark advance) position. The spark will stay at the 35 degree position until the airflow through the carburetor reaches a value corresponding to about 800 (approx. 2000 rpm) horsepower, when the spark returns to the 20 degree position. For automatic spark check, see Section II, paragraph 7.*c.*(10).

(5) PROPELLER GOVERNOR CONTROL.—The propeller governor control is located on the engine control unit. Move the control forward to increase rpm; move the control aft to decrease rpm. Full forward position gives take-off rpm (2800); full aft position gives approximately 1200 rpm. The control can be manipulated together with the throttle at or above maximum cruising

powers, thus obtaining a smooth coordination of manifold pressure, horsepower, and rpm. The control sets the constant speed unit and has no direct control over propeller blade angle. The blade angle is such that 2800 rpm can be obtained with slightly less than full power. Rapid changes in throttle or propeller control setting will tend to cause rpm to "overshoot the mark" momentarily before settling down.

CAUTION

The overspeed rating for this engine is 3120 rpm, for 30 seconds only.

(6) SUPERCHARGER CONTROL. — The control for the two stage, two speed supercharger is located on the engine control unit. The main stage impeller is driven directly by the crankshaft, and the auxiliary stage impeller is driven by two hydraulic couplings by means of which it can be engaged in either of two fixed gear ratios ("LOW" or "HIGH"). When the supercharger control is in the "NEUTRAL" position the auxiliary stage impeller is not in operation.

(a) The purpose of the auxiliary stage impeller is to supply air to the carburetor at approximately sea level pressure when operating at altitude. The auxiliary stage supercharger regulator maintains this condition by gradually opening the auxiliary stage gate valves, at the entrance to the auxiliary supercharger, as the altitude increases.

(b) The heat produced by compressing the intake air in the auxiliary stage supercharger is partially dissipated in the intercoolers, reducing the temperature of the supercharged air before it enters the carburetor. Refer to paragraph 2.b.(11) of this section.

(7) STARTER SWITCH.—The switch for the electric starter is located on the electrical control panel on the right hand shelf. The switch has two positions, "ON" and "OFF." A guard holds it in the "OFF" position when not in use.

(8) IGNITION SWITCH.—The switch for the ignition system is located on the left hand side of the instrument panel. It has four positions: "LEFT," "RIGHT," "BOTH" and "OFF."

(9) CARBURETOR AIR TEMPERATURE.—A warning light is provided on the main instrument panel to indicate (red light on) if the carburetor air temperature exceeds the maximum allowable limit of 43°C. (110°F.) Operating the engine at high power with excessively high carburetor air temperature (red light on) will probably cause detonation and serious damage to the engine except when operating at combat power. Detonation may be indicated by a slight undue roughness increasing somewhat in severity in a few seconds.

Note

When operating at combat power (water injection system "ON"), the carburetor air temperature warning light may be on without fear of detonation and possible damage to the engine. The light will not always come on, however, particularly at high altitudes when operating at combat power.

(a) Control of the carburetor air temperature is provided by means of the intercooler flap (see paragraph 2.b.(11), this section). This flap has an automatic control which is designed to give adequate cooling of the engine air for all normal operating conditions.

(b) The flap will move automatically to approximately the following settings:

  1. Normal climb or maneuvers—½ open.
  2. Severe operating conditions and maneuvers at low air speeds—full open.

(c) The above settings are to be set manually by means of the control switch on the center control panel in the event that the automatic action fails.

Note

If the warning light comes on when operating in low or high blower at low speeds, immediately open the intercooler flap wide by turning the intercooler flap switch to "OPEN."

(d) Excessively high carburetor air temperatures are most likely to occur during high power, high rpm, low air speed operation (as in steep climb or in tight turns) with the intercooler flap closed. The warning light is especially likely to come on if the supercharger control is shifted to a higher blower ratio at too low an altitude. In this case, immediately shift back to the next lower blower ratio.

(e) In view of the above, the warning light is placarded as follows: "SHIFT TO LOWER BLOWER IF LIGHT IS ON." This is the best and most positive means of reducing the carburetor air temperature to normal.

(10) COWL FLAP CONTROL.—The cowl flaps regulate the engine cooling, and are used to maintain the cylinder head temperature within the operating limits (200 to 245°C., 392 to 473°F.).

WARNING

CYLINDER HEAD TEMPERATURE MUST NOT EXCEED 245°C. (473°F.).

(a) The cowl flaps are controlled by a three-position switch located on the left hand shelf (see figure 10). The positions are: "OFF," "OPEN" and "CLOSE." The switching action for the "OPEN" and "CLOSE" positions is momentary, viz., the switch must be held in the selected position until the flaps reach the desired setting. No cowl flap position indicator is provided since the flaps are visible from the cockpit.

(b) The cowl flaps are actuated by a hydraulic strut which is controlled by a solenoid selector valve. The flaps are spring-loaded to open, their motion being restrained by a draw string cable connected to the hydraulic strut. A relief valve is provided to allow the flaps to "blow open" at indicated air speeds over 370 knots. This relieves the internal air pressure on the cowl panels.

(c) In general, for adequate cooling, the position of the cowl flaps for various flight conditions is as follows:

   1. Climb—open as required to maintain cylinder head temperature below 245°C. (473°F.).
   2. Level flight—"CLOSED."
   3. Diving—"CLOSED."

(11) INTERCOOLER FLAP CONTROL.

(a) The automatic intercooler flap control consists of an automatic pressure switch controlling the operation of a four-way, solenoid-hydraulic selector valve which is used to position the intercooler flap. The pressure switch maintains a constant cooling air pressure drop across the intercoolers. Since the intercoolers are not used when operating in neutral blower, an electrical override switch has been installed on the supercharger control handle which closes the intercooler flap when the control handle is in "NEUTRAL." On later airplanes a switch is incorporated which closes the intercooler flap at engine speeds below 2300 rpm in all blowers.

(b) The intercooler flap control switch, located on the center control panel, has four positions, "AUTOMATIC," "OFF," "OPEN," and "CLOSED." The "OPEN" and "CLOSED" positions are electrical overrides which are separate from the automatic control circuit and run directly to the solenoid valve. The switching action is momentary. The "OPEN" and "CLOSED" positions should be used only in the event that automatic action fails, or when it becomes desirable to cool the accessory compartment during ground operations. For all normal operations the switch should be turned to "AUTOMATIC."

(c) A relief valve is incorporated in the system to maintain flap operation within the strength limitations. This relief valve will have the same setting as the main relief valve of the hydraulic system; thus the flap will always "blow open" under the identical high speed air loads whether the control remains on "AUTOMATIC" or "OFF." However, this condition will occur only in a high speed dive.

(d) An adjustable hydraulic restrictor in the pressure line to the strut regulates the speed of flap travel. This restrictor should be so adjusted that travel from full closed to full open takes from 10 to 15 seconds. The position of the flap is shown by an indicator on the center control panel.

(12) OIL COOLER FLAP CONTROL.

(a) The oil cooler flaps are operated by a hydraulic selector valve which is thermally responsive to the return oil temperature. This control automatically positions the flaps so that they will begin to open from full closed when the oil reaches a temperature of 75°C. (167°F.) and will attain the full open position at 95°C. (203°F.). The control does not necessarily maintain a constant oil temperature but modulates through the above-mentioned temperature range.

(b) The cockpit control for the oil cooler flaps is a switch located on the center control panel. The switch has three positions: "OPEN," "CLOSED," "AUTOMATIC." The first two positions are electrical override controls for manual operation and should be used only when automatic action fails. In normal operation the position of the control is "AUTOMATIC." The "AUTOMATIC" position of the control is actually the off position of the switch since the automatic operation is purely hydraulic-mechanical and requires no electrical actuating impulse.

c. FUEL SYSTEM CONTROLS.

(1) GENERAL.—The fuel system controls are shown on figure 8.

   (a) FUEL.—Grade: 100/130.
       Specification: AN-F-28.

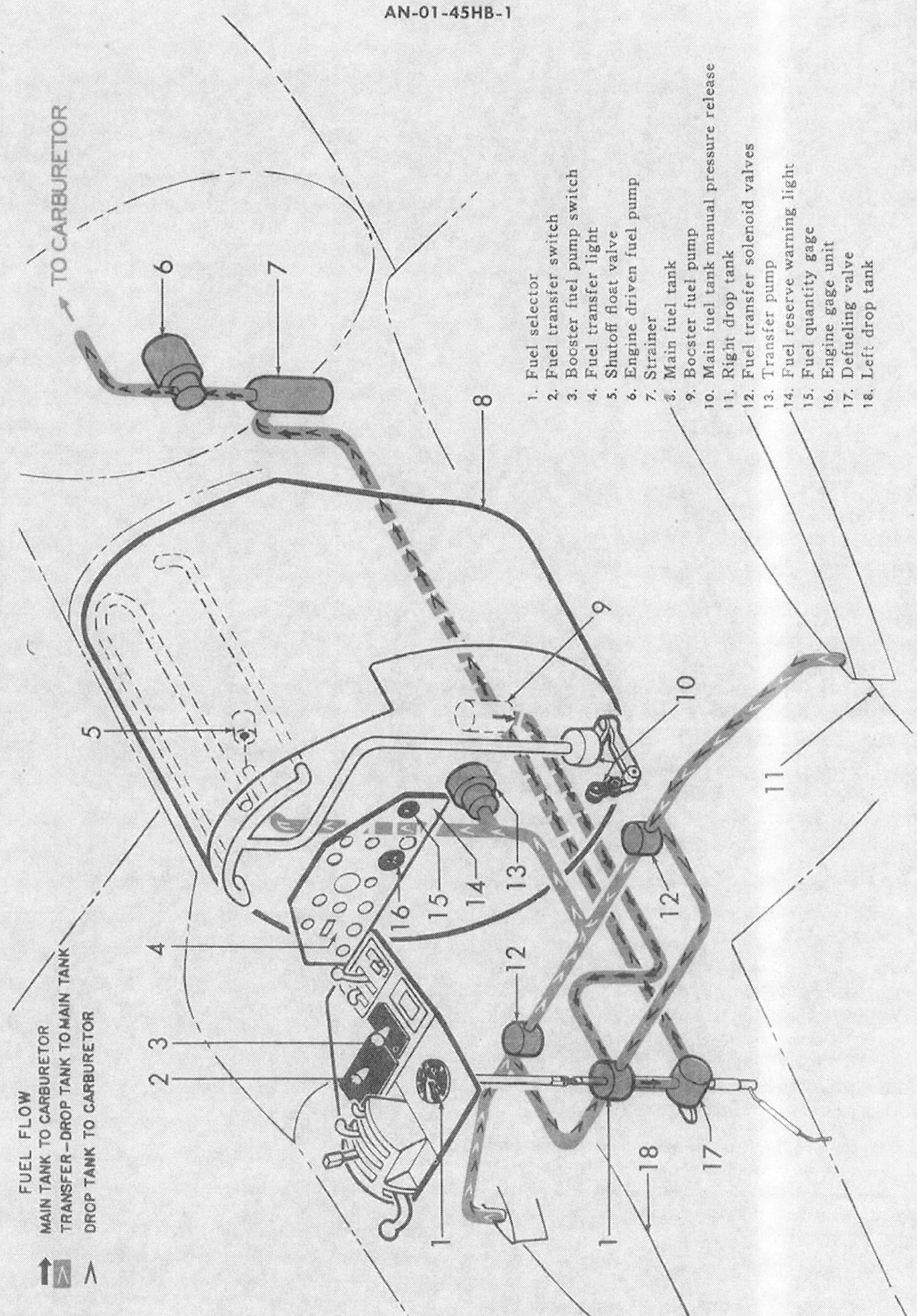

Section 1

AN-01-45HB-1

FUEL FLOW
- MAIN TANK TO CARBURETOR
- TRANSFER—DROP TANK TO MAIN TANK
- DROP TANK TO CARBURETOR

1. Fuel selector
2. Fuel transfer switch
3. Booster fuel pump switch
4. Fuel transfer light
5. Shutoff float valve
6. Engine driven fuel pump
7. Strainer
8. Main fuel tank
9. Booster fuel pump
10. Main fuel tank manual pressure release
11. Right drop tank
12. Fuel transfer solenoid valves
13. Transfer pump
14. Fuel reserve warning light
15. Fuel quantity gage
16. Engine gage unit
17. Defueling valve
18. Left drop tank

Figure 8—Fuel System Control Diagram

(2) TANKS.

(a) The self-sealing main tank, located in the fuselage, forward of the cockpit, has a capacity of 233 U. S. gallons (194 Imp. gallons) of fuel. A red warning light is provided on the instrument panel to indicate when approximately 50 U. S. gallons (42 Imp. gallons) of fuel or less remain in the main tank.

(b) Provision is made, on the center section twin pylons, for carrying Navy Standard 150 U.S. gallon (125 Imp. gallons) droppable tanks, either self-sealing or non-self-sealing, or Navy Standard self-sealing 100 U.S. gallon (83 Imp. gallon) droppable tanks.

Note

P-38 steel droppable tanks of 171 U.S. gallon (142 Imp. gallon) fuel capacity or P-38 self-sealing droppable tanks of 159 U.S. gallon (132 Imp. gallon) capacity may be substituted for Navy Standard tanks when the latter are not available.

(3) FUEL QUANTITY GAGE.—The electric fuel quantity gage, located on the instrument panel, is calibrated to indicate correctly the quantity of fuel in the main tank with the airplane in level flight at approximately 180 knots indicated air speed, normal fighter load.

(4) FUEL TRANSFER SWITCH. — A three-position switch, located on the left hand control shelf (see figure 10), controls the transfer pump and the solenoid valves in the transfer lines at the same time; the positions of the switch are: "LEFT," "RIGHT," "OFF." Refer to Section II, paragraph 4.e. for information concerning transfer pump operation.

(5) BOOSTER FUEL PUMP SWITCH.—The booster fuel pump is operated by a three-position switch located on the left hand shelf (see figure 11). The three positions of the switch are: "BOOST," "EMERGENCY," "OFF." Refer to Section II, paragraph 4.g. for information concerning booster fuel pump operation.

(6) FUEL TANK SELECTOR.—The fuel tank selector is located on the left hand control shelf (see figure 23). It has four positions: "ON," "OFF," "RIGHT DROP STANDBY," "LEFT DROP STANDBY."

d. OIL SYSTEM CONTROLS.

(1) GENERAL.—The oil system controls are shown on figure 9.

(2) OIL.—Grade: 1100
Specification: AN-VV-O-446.

(3) OIL TANK.—The oil tank, located just forward of the firewall, has a capacity of 18½ U.S. gallons (15.4 Imp. gallons) of oil. The warm-up compartment of the tank is equipped with a pendulum which supplies oil to the engine under all flight conditions.

WARNING

Only enough oil is available for 10 seconds inverted flight.

(4) OIL COOLERS. — The oil coolers, mounted just aft of the air duct openings in the wings, are provided with a thermostatic regulator valve set to maintain the oil temperature at 75° to 95°C. (167° to 203°F.). This regulator valve operates the oil cooler air exit flaps located under and just aft of each oil cooler. In addition each cooler is provided with a temperature regulator valve which regulates the oil flow through the internal passages of the oil cooler.

(5) OIL COOLER SHUTOFF CONTROLS.—The oil system is equipped with a shutoff valve by means of which the pilot can cut out either oil cooler at his discretion. The valve is operated by two handles located on the center control panel. In order to shut off an oil cooler, pull the corresponding handle and turn in a clockwise direction. Both handles shall be forward for normal operation.

Note

If one of the coolers is damaged, such damage will not be indicated on either the oil pressure gage or oil temperature gage since these values will not change appreciably until all of the oil is out of the system. Therefore, during combat conditions, the pilot should check both oil coolers frequently at the trailing edge of both wings, aft of the oil coolers, to see if there is any oil leakage. If oil leakage is visible, by-pass the damaged oil cooler by turning the proper handle to the shut-off position.

(6) OIL DILUTION SYSTEM. — The oil dilution switch is located on the right hand shelf. The purpose of the oil dilution system is to reduce the cranking torque of the engine and to provide a sufficient amount of low viscosity oil for lubrication when starting the engine at temperatures near or below the pour point of the oil. This is accomplished by diluting the oil in the engine and

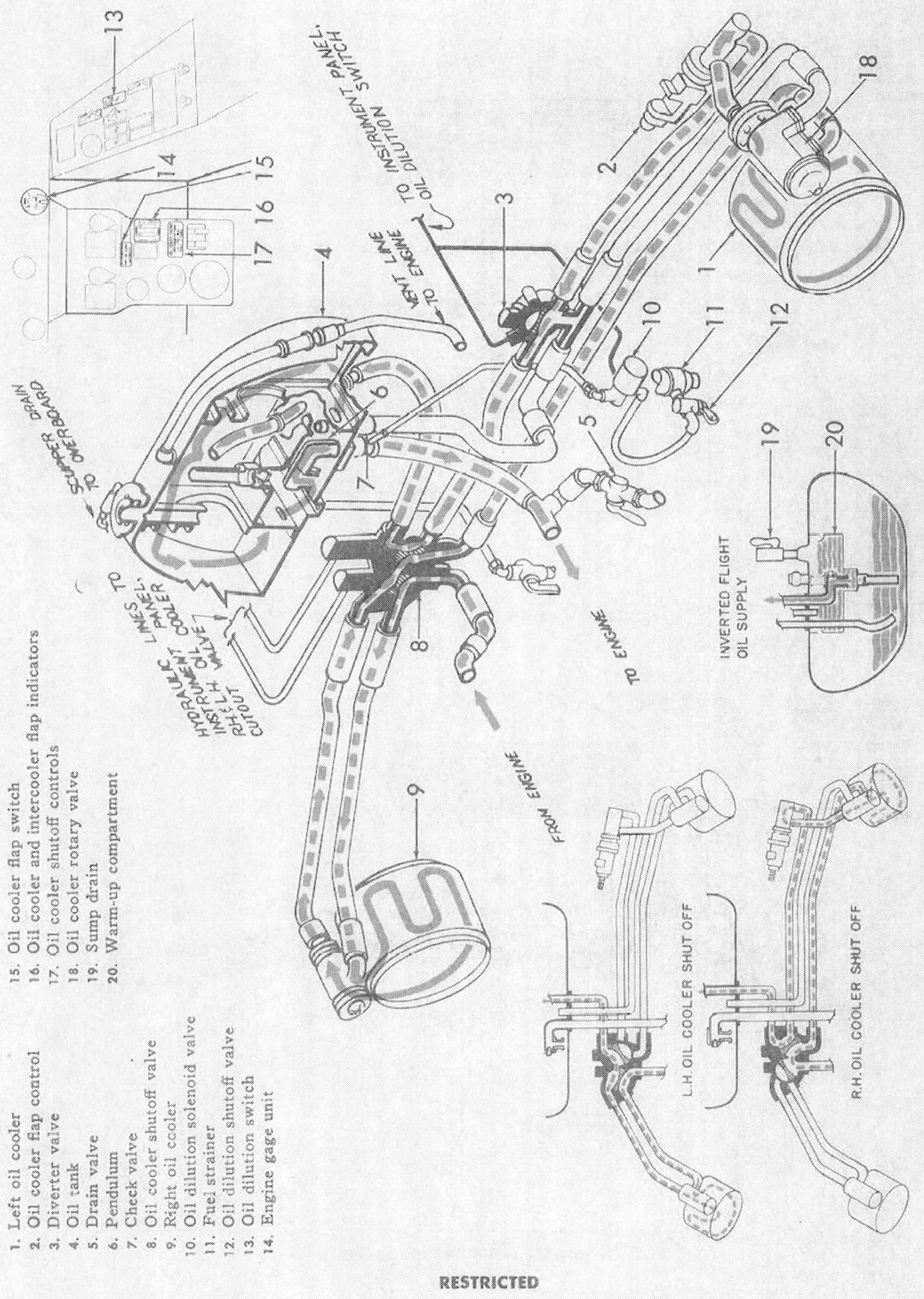
1. Left oil cooler
2. Oil cooler flap control
3. Diverter valve
4. Oil tank
5. Drain valve
6. Pendulum
7. Check valve
8. Oil cooler shutoff valve
9. Right oil cooler
10. Oil dilution solenoid valve
11. Fuel strainer
12. Oil dilution shutoff valve
13. Oil dilution switch
14. Engine gage unit
15. Oil cooler flap switch
16. Oil cooler and intercooler flap indicators
17. Oil cooler shutoff controls
18. Oil cooler rotary valve
19. Sump drain
20. Warm-up compartment
SCUPPER DRAIN TO OVERBOARD
VENT LINE TO ENGINE
TO INSTRUMENT PANEL, OIL DILUTION SWITCH
TO LINES, PANEL, COOLER
HYDRAULIC INSTRUMENT OIL INSTR. L.H. VALVE R.H. CUTOUT
TO ENGINE
FROM ENGINE
INVERTED FLIGHT OIL SUPPLY
L.H. OIL COOLER SHUT OFF
R.H. OIL COOLER SHUT OFF

Figure 9 — Oil System Control Diagram

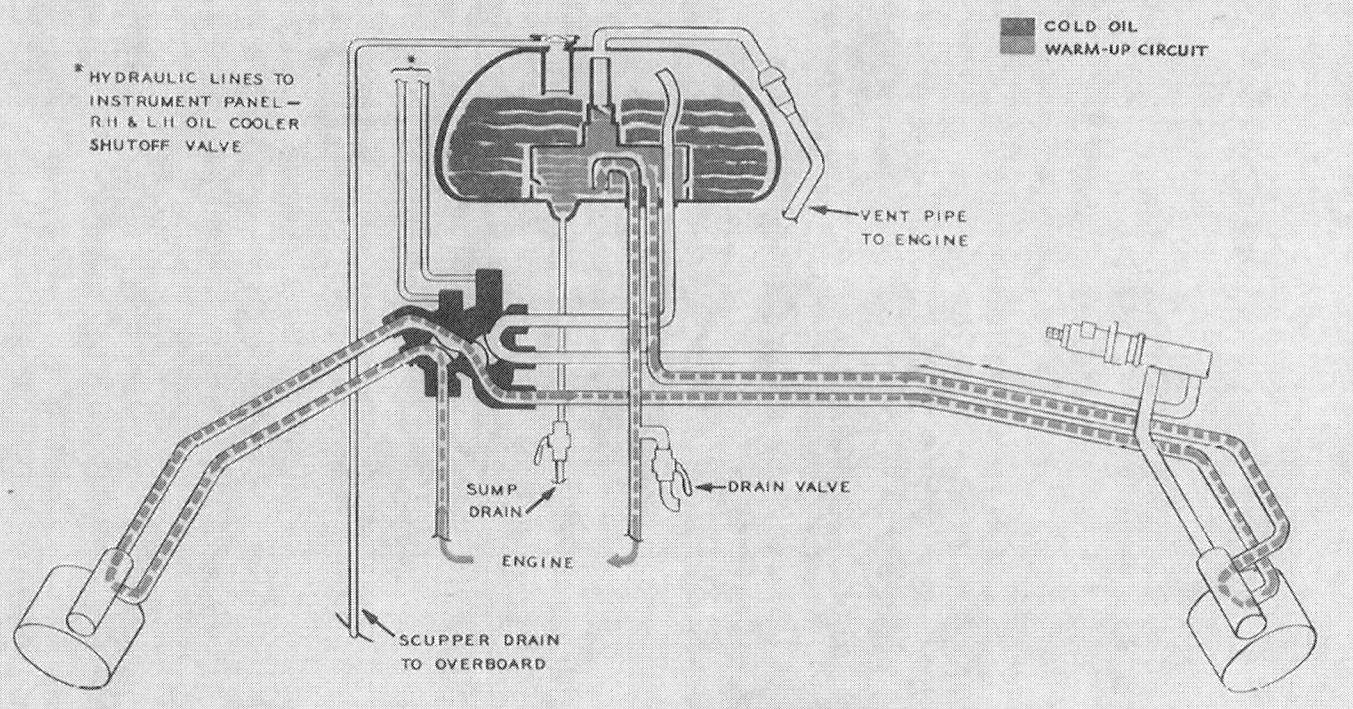

Figure 9A — Oil System Warm-Up Circuit

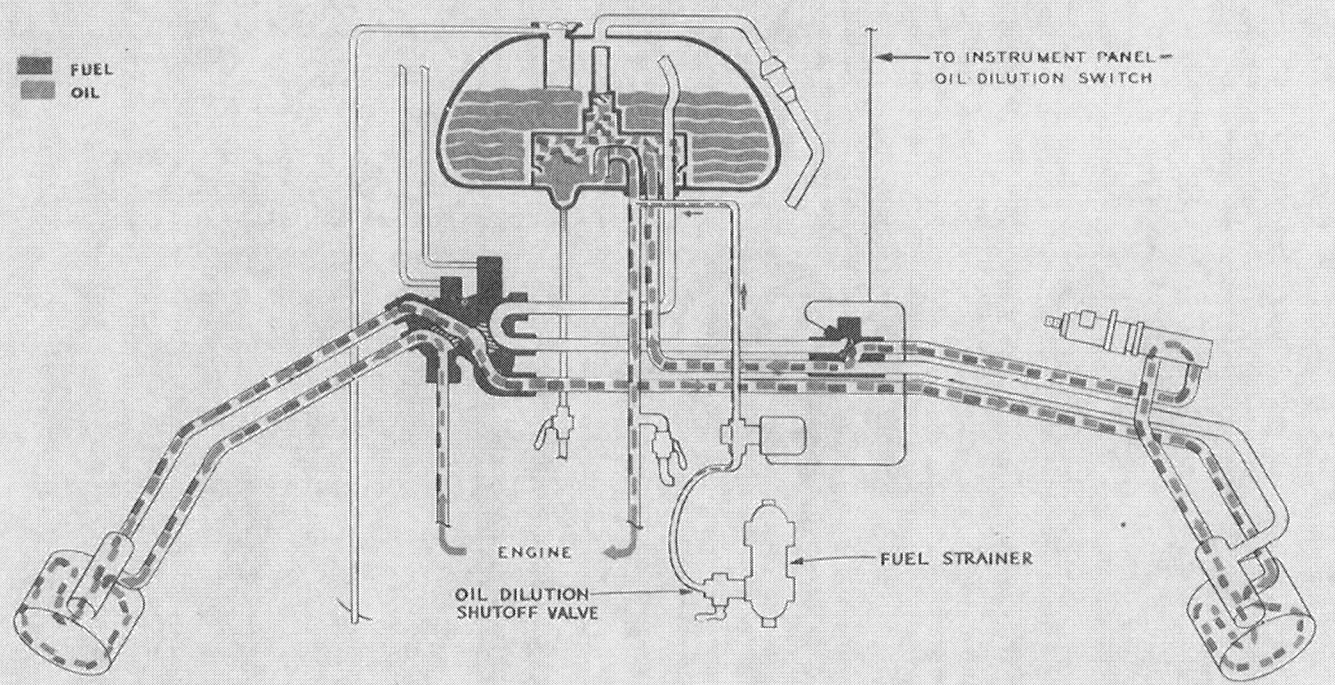

Figure 9B — Oil Dilution System

quick warm-up circuit of the oil system with gasoline as is described in Section II, paragraph 21.*d*. Oil dilution can be effected only when the engine is running, and is done just prior to stopping the engine in preparation for the next start. Never dilute subsequent to starting a cold engine having undiluted oil as it will be of no aid in starting and may cause trouble.

Note

THE OIL DILUTION SYSTEM IS NOT INSTALLED IN THE AIRPLANE AT DELIVERY. ONE OIL DILUTION KIT, CONTAINING THE COMPLETE SYSTEM, IS FURNISHED WITH EVERY TENTH AIRPLANE, FOR INSTALLATION IN EXTREME COLD WEATHER CONDITIONS.

e. HYDRAULIC SYSTEM CONTROLS.— The hydraulic system controls are shown on figures 10, 12, 13, 14, 15, 16, 17, 18 and 19.

Note

The cooling flaps controls are discussed in paragraphs 2.*b*.(10), (11), and (12), this section.

(1) HYDRAULIC OIL: Specification: AN-VV-O-366 (red fluid).

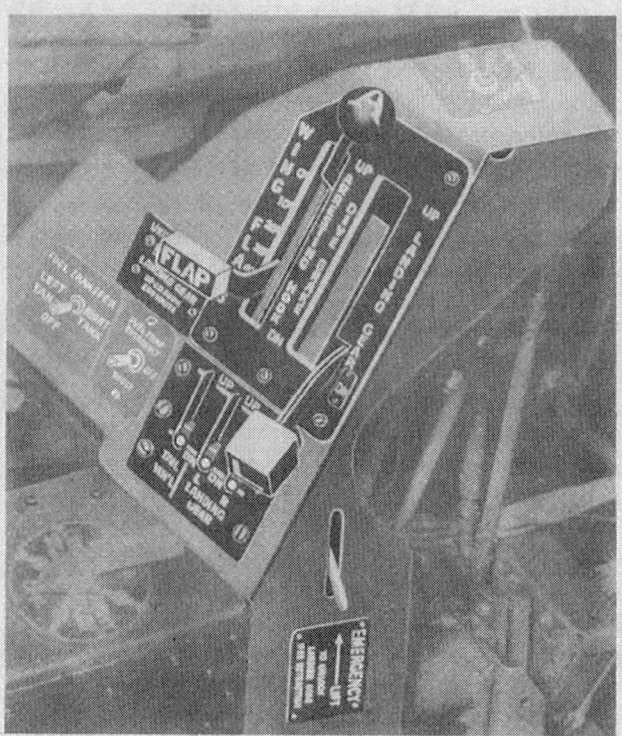

Figure 10—Inclined Panel — Left Hand Shelf

(2) GENERAL.—An engine-driven hydraulic pump, pressure regulator and accumulator combine to maintain a constant pressure of 900 to 1150 pounds per square inch in flight, as indicated by the gage on the center control panel in the cockpit. After a hydraulic control is moved, the pressure will drop and vary while the units are moving, and then become steady after the movement is completed. A hand pump is provided for use when the engine is not running or in the event of failure of the engine-driven hydraulic pump.

(3) LANDING GEAR CONTROL. (See figure 13).

(*a*) The control for landing gear retraction and extension is on the left hand control shelf. To operate the landing gear, move the control to the desired position. The landing gear and closure doors operate automatically in proper sequence. The positions of each side of the landing gear and of the tail wheel are shown by the respective indicators located on the left hand shelf.

(*b*) MANUAL EMERGENCY OVERRIDE.—The landing gear control is electrically locked in the "DOWN" position and cannot be moved out of that position as long as the weight of the airplane is supported by the landing gear. A manual emergency override is installed on the left hand shelf beside the landing gear indicators (see figure 11). If the electrical lock in the landing gear fails to operate after take-off (because of an electrical or mechanical failure), lift the manual override to release the landing gear control.

(*c*) The landing gear and arresting hook are mechanically interlocked so that the arresting hook cannot be lowered unless the landing gear is lowered. Conversely, the landing gear cannot be retracted without retracting the arresting hook.

(4) DIVE BRAKE CONTROL. (See Figure 14.)

(*a*) The dive brake control is located on the left hand shelf. To extend the dive brake, move the landing gear control into the "DOWN" position of the dive brake slot; this extends the main landing gear only, leaving the tail wheel retracted. To retract the dive brake, move the control to "UP."

(5) ARRESTING HOOK CONTROL. (See figure 15.)—The control for the arresting hook is located on the left hand shelf. The control has two positions, viz., "UP" and "DOWN," and must be moved to the limit stops of these positions for the desired action of the arresting hook.

(6) WING FLAP CONTROL. (See figure 16.)

(a) The wing flap control mechanism is designed so that any flap angle in 10 degree steps to full "DOWN" (50 degrees) can be obtained by a corresponding setting of the wing flap control located on the vertical panel of the left hand shelf. Refer to Section IV, paragraph 3, for information concerning emergency wing flap operation.

(b) The wing flap system includes a mechanism which causes the flaps to "blow up" (back off) from the angle set by the control under excessive air loads caused by air speeds greater than normal. The flaps will return to the angle corresponding to the control setting when the air speed is reduced. The mechanism is set so that with flaps full down (50 degrees), and power on for level flight in the landing condition, they begin to "blow up" between 100 and 115 knots indicated. At lesser flap settings, the "blow-up" speeds will be greater than with flaps full down.

Note

The wing flap control shall not be placed in position for lowering flaps at speeds in excess of 200 knots even though the flaps are protected by an overload relief mechanism. If the flap relief mechanism is not in operation, the restricted speed with flaps down varies from 130 knots, with flaps deflected 50 degrees, to 200 knots with flaps deflected 20 degrees.

(c) The flaps are also designed for use in maneuvering the airplane in combat. With typical maneuvering flap deflections of 20 degrees or less the airplane may be maneuvered at equivalent limiting "flaps up" accelerations up to 200 knots.

(7) WING FOLDING AND LOCKING CONTROLS. (See figure 17.)

(a) The wing folding control, located on the left hand control shelf, has the following positions: "SPREAD," "NEUTRAL," "FOLD."
The wing hinge pin locking control is adjacent to the wing folding control.

(b) To fold the wings, release the latch holding the locking control in position and move the locking control to "UNLOCKED." When the latch is released and the locking control is moved aft, the wing hinge pin locks are relieved and the pin is freed for extraction. Placing the folding control in the "FOLD" position extracts the wing hinge pins and folds the wings in the proper sequence. When the operation is complete, move the folding control to "NEUTRAL." The wings may be folded or spread manually by means of the hand pump whenever the engine is not running.

(c) To spread the wings, move the wing folding control to "SPREAD." This action moves the wings to their complete spread position and inserts the hinge pins in proper sequence. When this operation is complete, move the locking control forward and see that the latch locks firmly behind the control lever. This action locks the wing hinge pins.

Note

For all flight operations, the wing folding control shall be in the "SPREAD" position, and the locking control shall be forward in its locked position.

(d) A visual check that the wings are fully spread and that the wing hinge pins are "home" is provided by the closing doors (painted red inside) at the wing joint. These doors will not close until the outer panels are completely spread and the wing hinge pins are "home." See figure 11.

1. A red warning flag has been provided at each wing folding joint to indicate when the wing hinge pin lock is not in position. The warning flag will drop flush with the wing surface only when the locking pin is actually in the locked position. See figure 11.

(e) No provision is made in the wing folding part of the hydraulic system to keep the outer panels "in step"; that is, no flow equalizer is installed. The wings must not be left free in any intermediate position between fully spread and fully folded, as air loads will cause them to shift positions, blowing one down and the other up. When fully folded, the wings should be locked by means of the jury struts. When the wings are fully folded and the jury struts are installed, the wings, may, by temporarily unlocking the jury struts, be moved to vertical for gun servicing by the action of the accumulator, if the pressure is

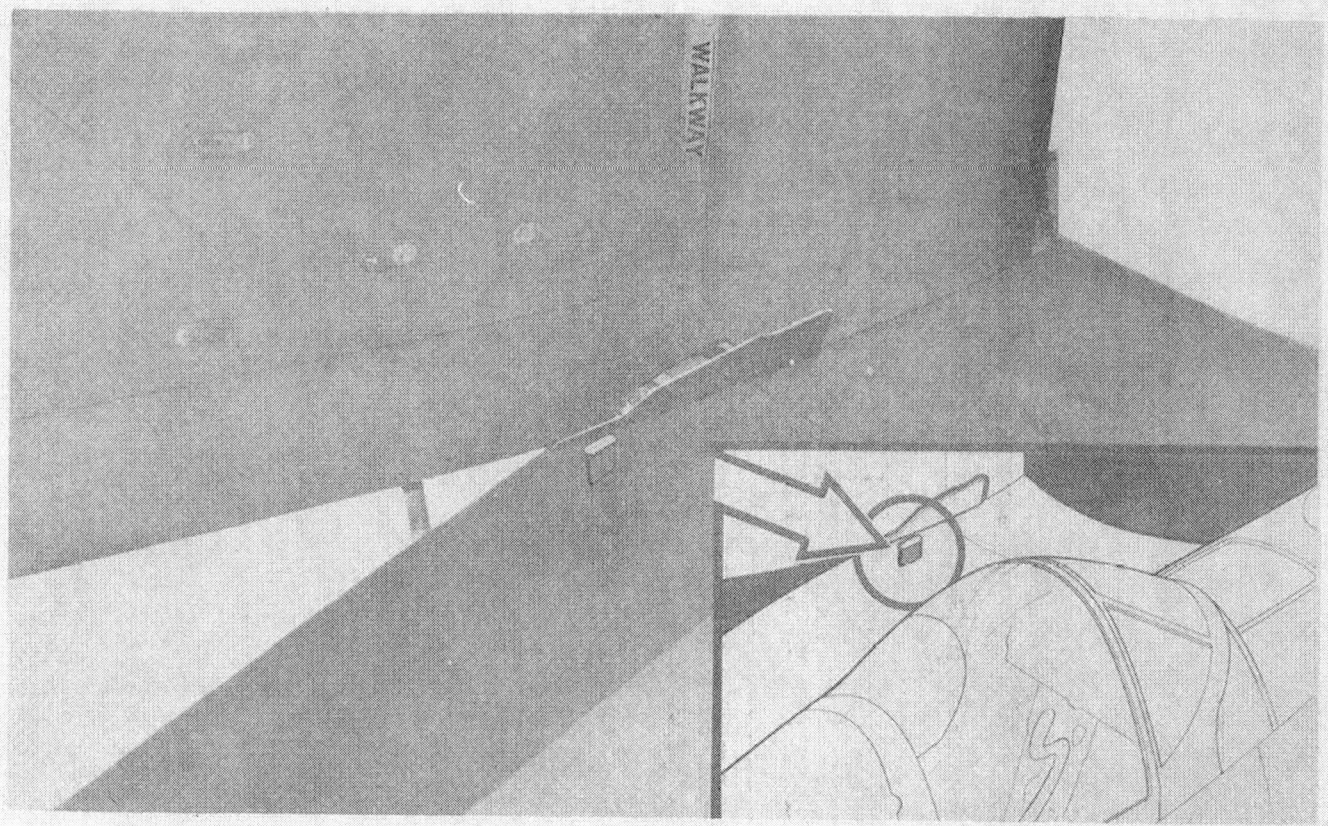

Figure 11 — Wing Lock Pin Indicator System

up, or by the hand pump. The jury struts are telescopic, with a limit stop at the vertical position. To fold only one wing, hold the opposite wing down (two or three men at the tip). By locking the "up" wing with a jury strut, the wings will remain in this position as long as the wing fold control is in "NEUTRAL."

(8) GUN CHARGING CONTROLS.

(a) The fixed guns are charged hydraulically. The two charging knobs are located on the center control panel. The left charging knob operates the safetying and charging of the left guns, and the right knob controls the right guns.

(b) To charge the guns, turn the knobs all the way over to "CHARGE." Be sure that the knobs are not stuck, thereby preventing a complete turn to "CHARGE." Push the knobs in all the way. Do not hold them in, as the continued pressure may cause them to stick. When the knobs spring out, the guns are charged.

(c) To safety the guns, turn the charging knobs to "SAFE" and push in. The knobs will spring out when the guns are safe. The gun chargers will then hold the bolts back in the "SAFE" position. To allow the bolts to go forward from "SAFE" to "CHARGE," simply turn the knobs to "CHARGE."

(d) Simply rotating the knob to "SAFE" does not safety the guns unless the knob is pushed in. No ammunition is lost in pushing the knob in while in the "SAFE" position and hence, if in any doubt, no harm is done by pushing the knob in again for the pilot's own reassurance.

CAUTION

Always safety the guns before landing.

(e) To clear jams during flight, charge the guns a few times. Ejected cartridges can be seen passing the trailing edge of the outer panel flap.

(9) HAND PUMP.—The feed for the hand pump is drawn from the bottom of the hydraulic reservoir, while that for the engine-driven pump is drawn from the ½ gallon level of the reservoir. If failure of a hydraulic pressure line allows all of the available fluid to be pumped overboard, the ½ gallon of hydraulic oil remaining in the reservoir is enough for one operation each by use of the hand pump, of the wing flaps, cooling flaps, and

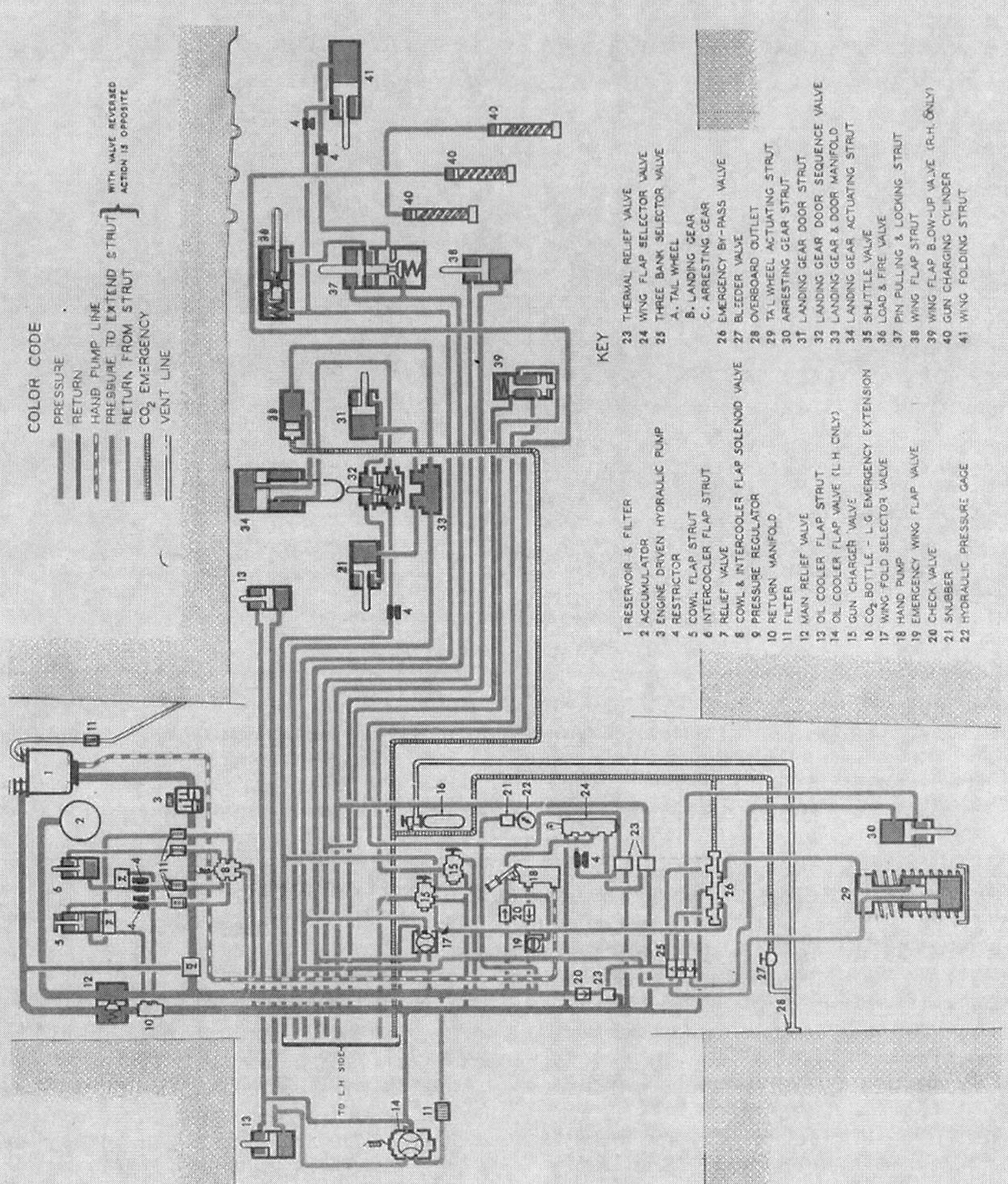

Figure 12 — Hydraulic System Schematic

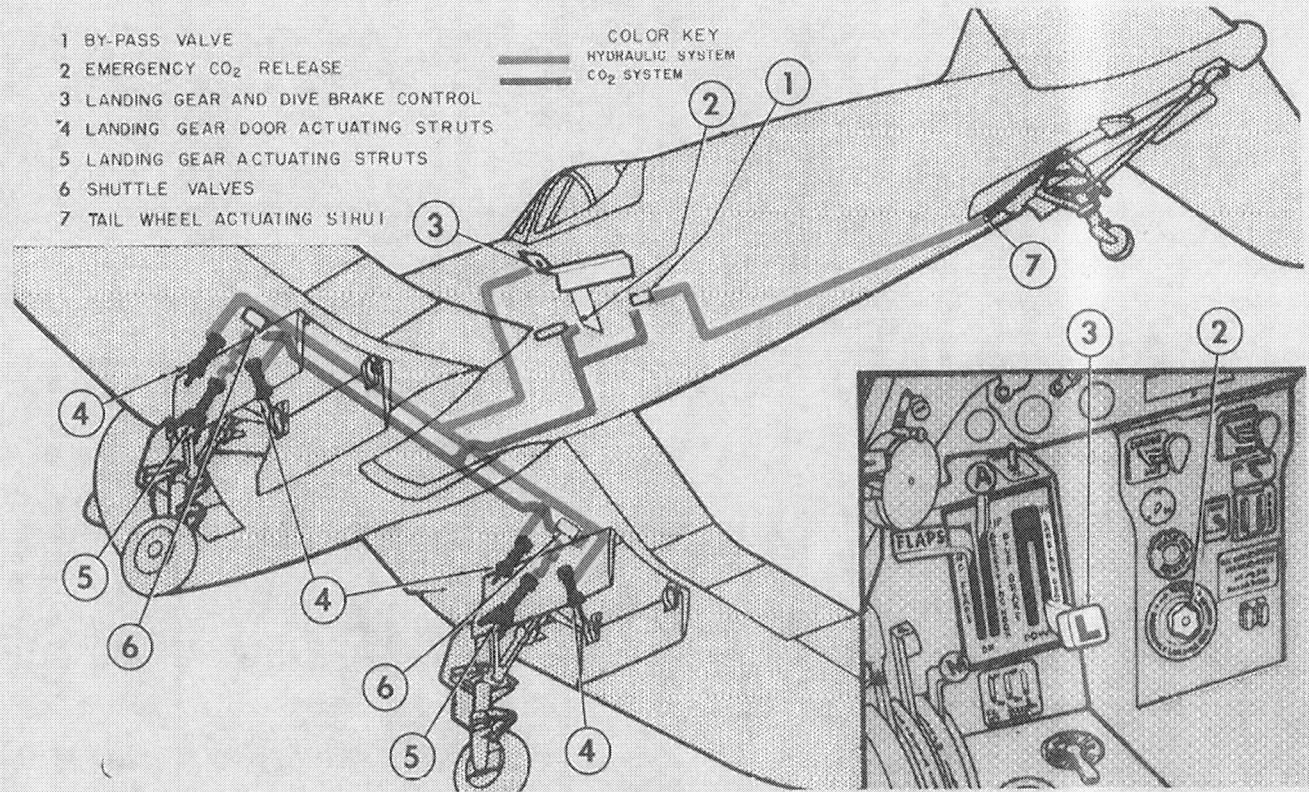

Figure 13 — *Landing Gear Hydraulic System Control Diagram*

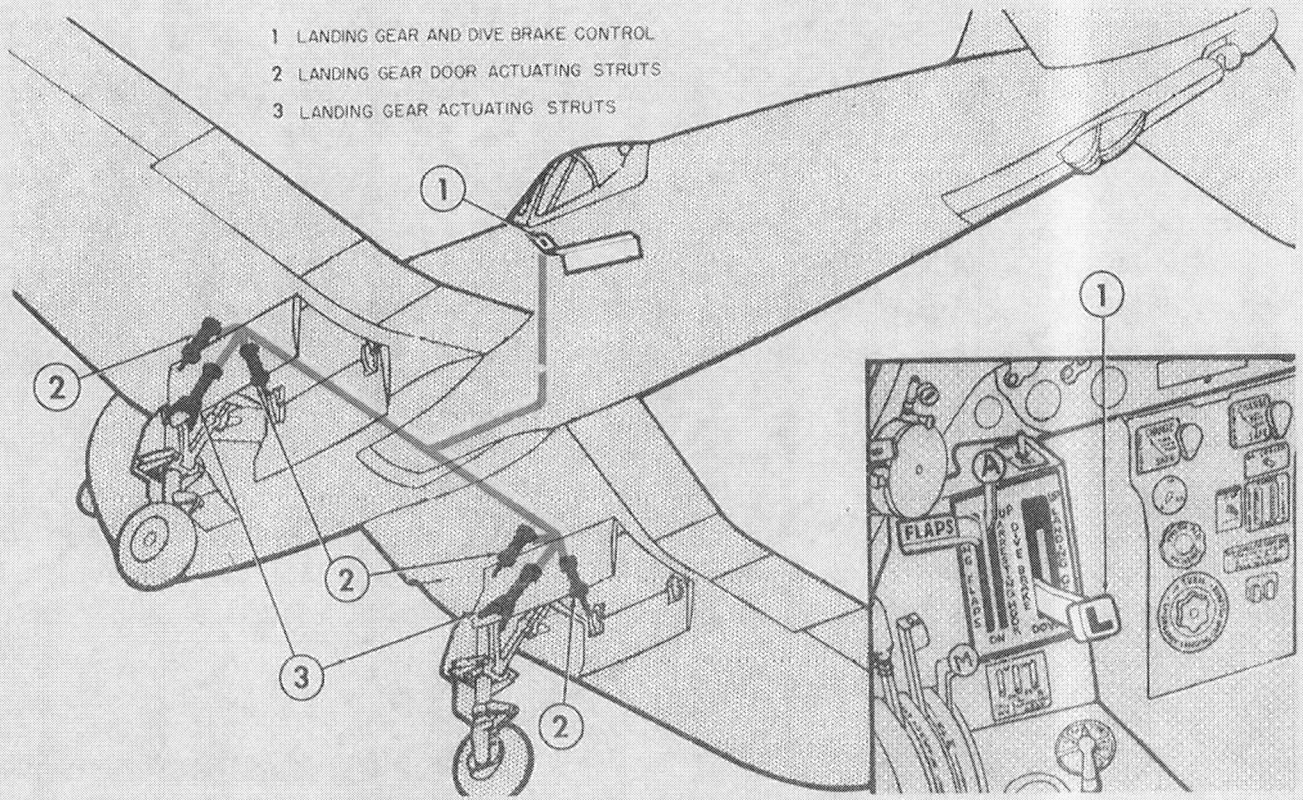

Figure 14 — *Dive Brake Hydraulic System Control Diagram*

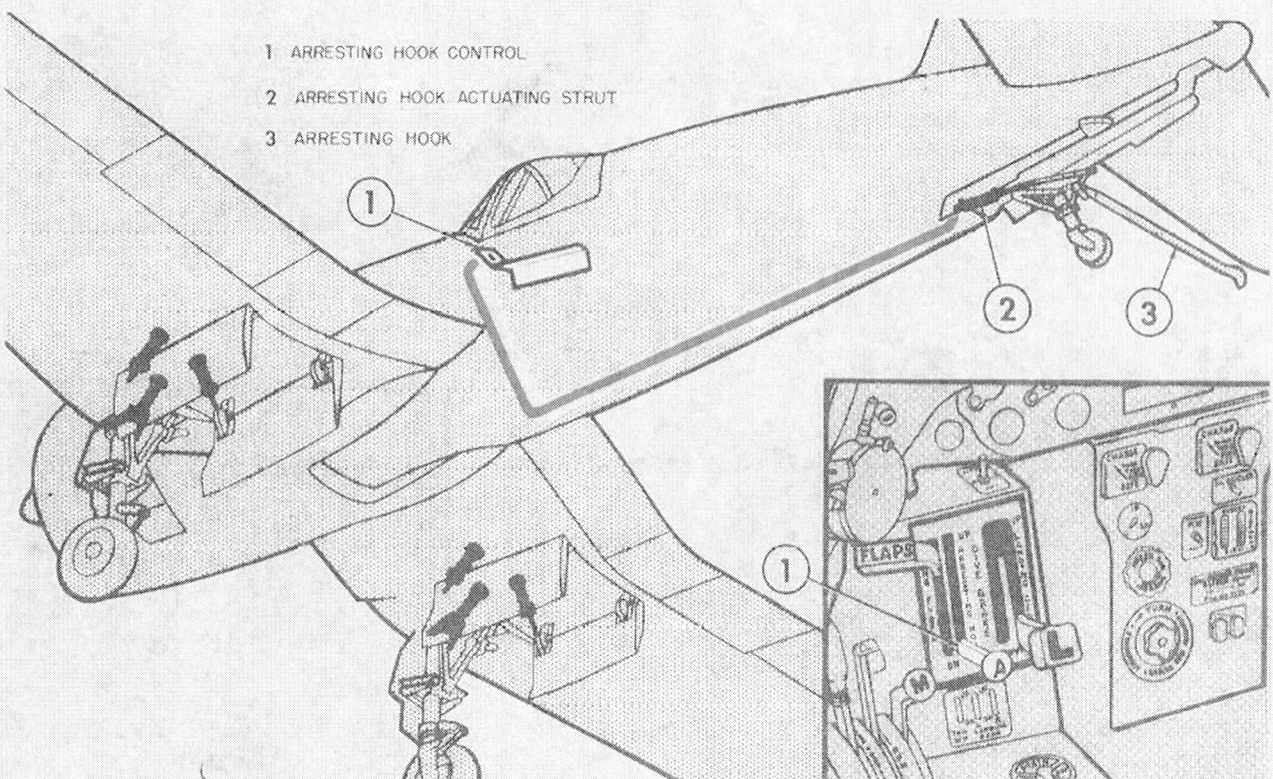

Figure 15 — Arresting Hook Hydraulic System Control Diagram

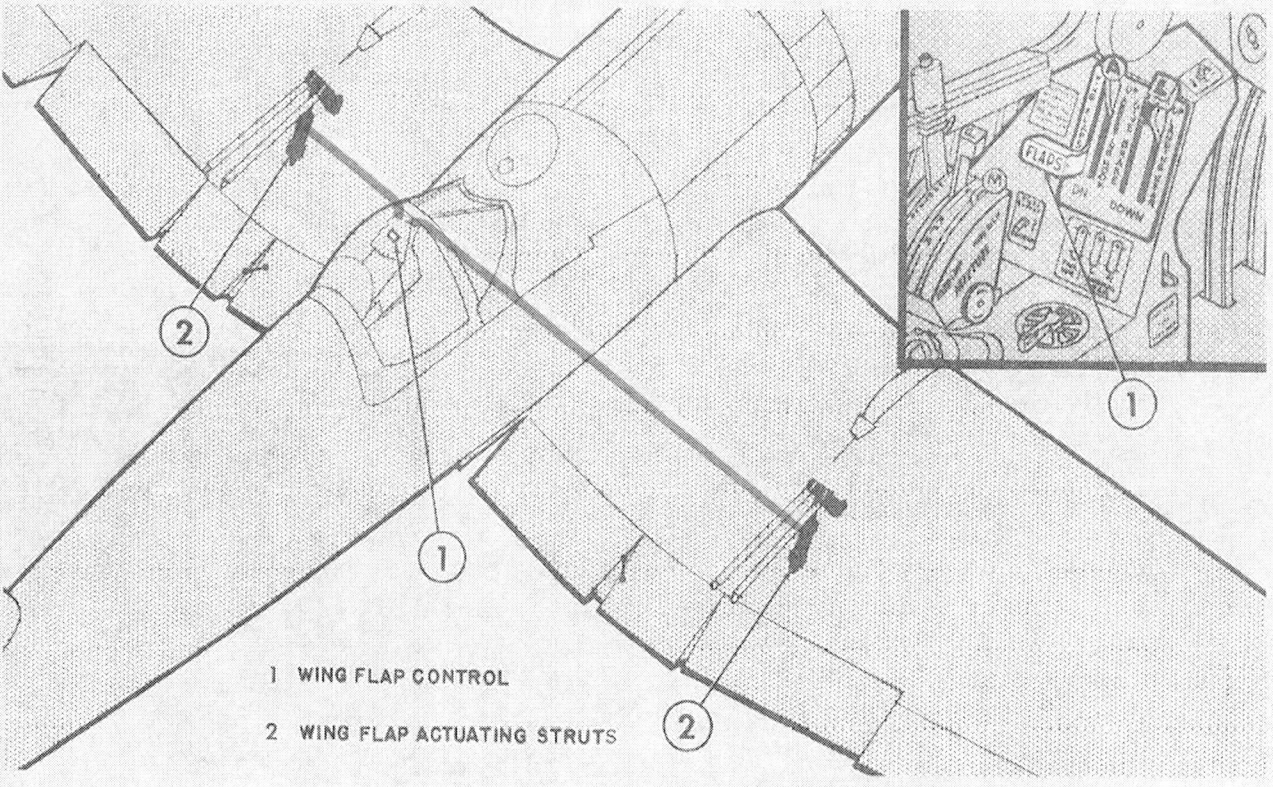

Figure 16 — Wing Flap Hydraulic System Control Diagram

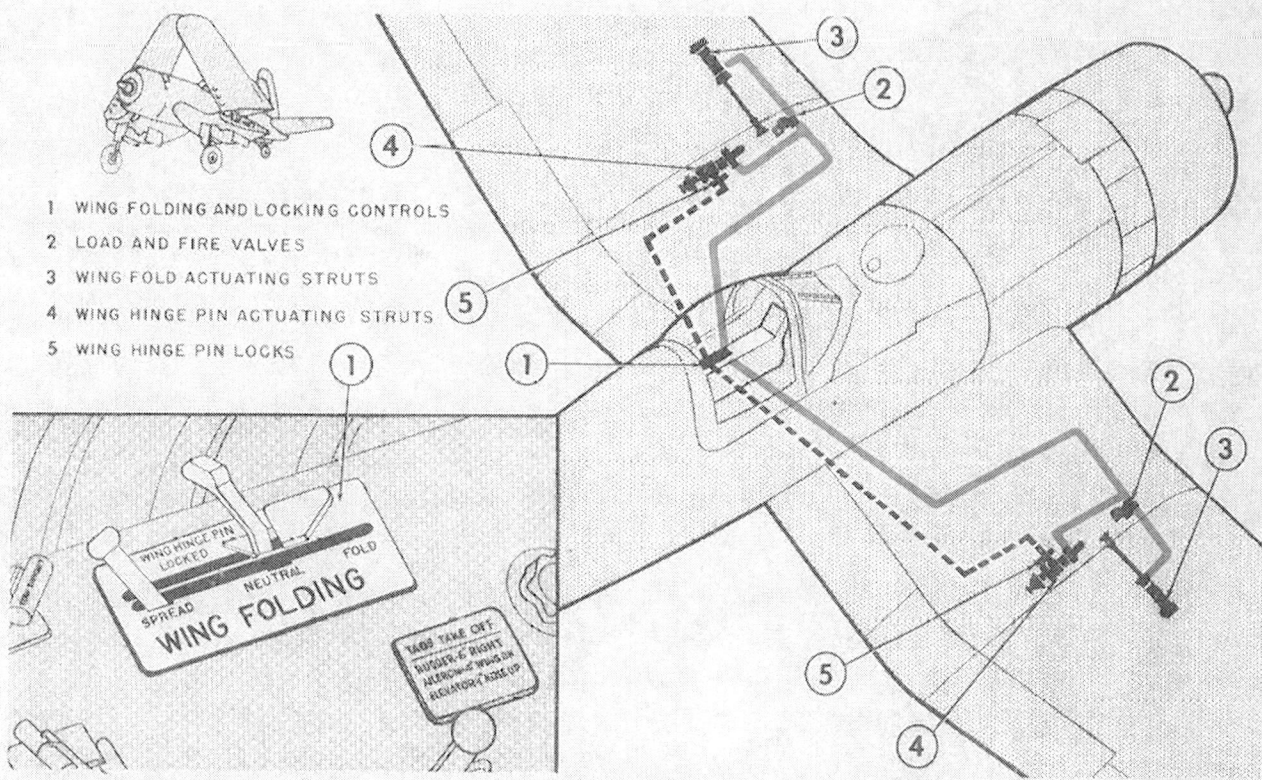

Figure 17 — Wing Folding Hydraulic System Control Diagram

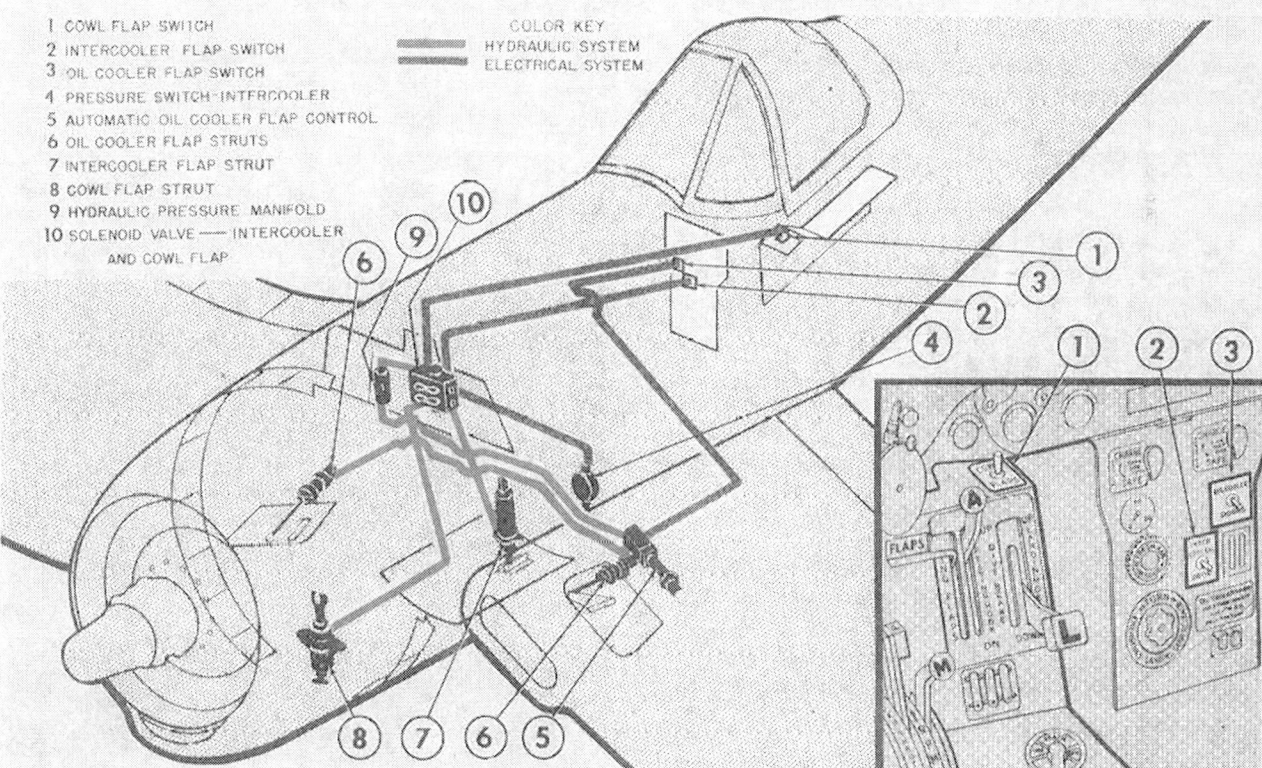

Figure 18 — Cooling Flaps Hydraulic System Control Diagram

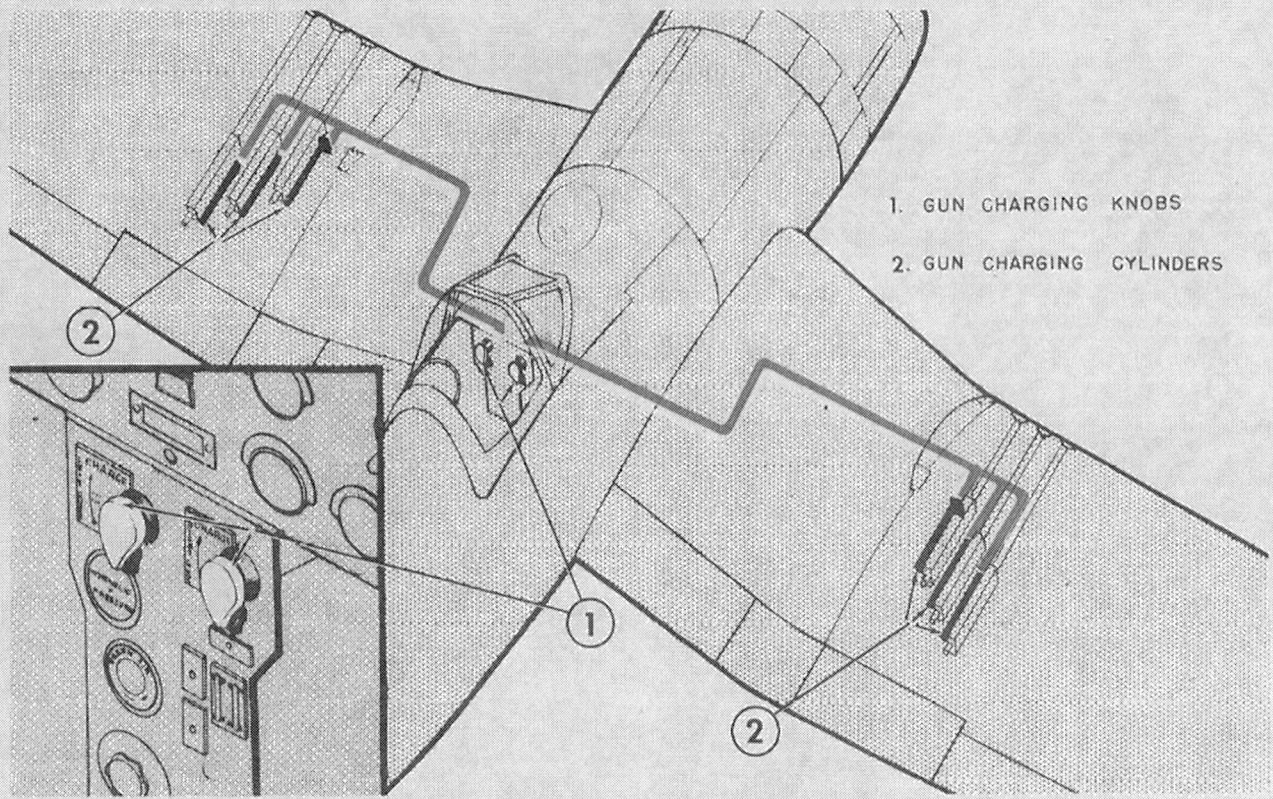

Figure 19 — Gun Charging Hydraulic System Control Diagram

gun charging. The arresting hook does not require hydraulic pressure for extension. Emergency landing gear extension is provided by the CO_2 extension system (see Section IV, paragraph 2).

Note

If it is known that the hydraulic system has lost fluid, the landing gear should be lowered by means of the emergency extension system, conserving the remaining fluid for lowering the wing flaps, etc.

f. TRIM TAB CONTROLS.—Trim tabs are provided on the left wing aileron, on the elevators, and on the rudder to permit control forces to be trimmed to comfortable values under all normal operating conditions. Refer to figure 20.

(1) AILERON TAB CONTROL.—Rotating the aileron tab control (inclined wheel on left hand shelf) to the right results in a downward movement of the right wing in fight. Rotating the hand wheel to the left results in upward movement of the right wing.

(2) ELEVATOR TAB CONTROL.—Rotating the elevator trim tab control (large vertical wheel on the side of the left hand shelf) forward lowers the nose of the airplane in flight. Aft rotation raises the nose.

(3) RUDDER TAB CONTROL.—Rotating the rudder tab control (horizontal hand wheel on left hand shelf) to the right moves the nose of the airplane to the right in flight. Rotating the

Figure 20 — Trim Tab Controls

hand wheel to the left moves the nose of the airplane to the left.

g. BALANCE TABS.—Balance tabs are provided on the ailerons and elevators in order to reduce the stick forces. These tabs require no control, since they are linked directly to the control surfaces. Down movement of the ailerons and elevators causes up movement of the tabs, and vice versa.

h. MISCELLANEOUS CONTROLS AND EQUIPMENT.

(1) GENERAL.—Other controls and equipment, not discussed below, are located in Sections IV and V, as follows:

(*a*) Emergency Egress—Section IV, paragraph 1.

(*b*) Life Raft—Section IV, paragraph 4.

(*c*) Emergency Landing Gear Operation—Section IV, paragraph 2.

(*d*) Emergency Wing Flap Operation—Section IV, paragraph 3.

(*e*) Operation of Oxygen Equipment—Section V, paragraph 1.

(*f*) Operation of Radio Equipment—Section V, paragraph 2.

(*g*) Operation of Electrical Equipment—Section V, paragraph 3.

(*h*) Operation of Armament — Section V, paragraph 4.

Figure 21 — Cabin Control

(2) COCKPIT CABIN CONTROL. — The cockpit cabin is opened and closed from the inside by means of a control handle located on the right hand side of the cockpit. External operation of the cabin is accomplished by means of a thumb button located on the right hand side panel of the windshield, just forward of the cabin opening. Additional details of the latter operation are given in Section II, paragraph 2. (See figures 4 and 21.)

(3) RUDDER PEDAL ADJUSTMENT.—A fore and aft ratchet-type adjustment of five inches is provided on each rudder pedal. To adjust a pedal for optimum comfort, press the toe release just outboard of the pedal and the pedal will move to its most forward position. Hook the foot under the selected pedal and pull it aft to the desired position. Adjust the other pedal in the same manner.

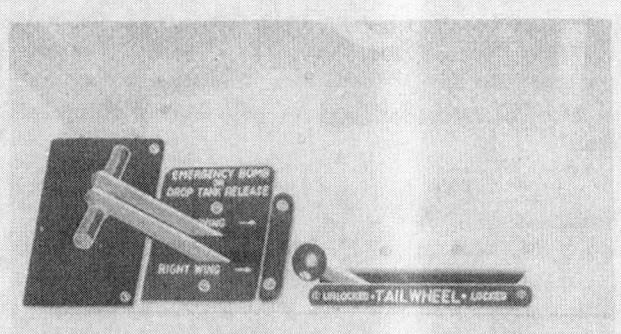

Figure 22 — Tail Wheel Lock and Manual Drop Tank Release

(4) TAIL WHEEL LOCK CONTROL.—The tail wheel lock control is located on the left hand shelf just aft of the rudder tab control. It has two positions, "LOCKED" and "UNLOCKED." To lock or unlock the tail wheel, move the control into the desired position. (See figure 22.)

(5) WINDSHIELD DEFROSTER CONTROL.

(*a*) The regulator control for defrosting the windshield is located on the instrument panel, directly beneath the bomb switch on the armament switch box. The control has two positions, "ON" and "OFF." The quantity of hot air necessary for adequate defrosting is regulated by turning the control to the desired position between "ON" and "OFF."

(*b*) Heated air for defrosting the windshield is taken from behind the right hand oil cooler and passed up to the cowl deck through a series of ducts.

Note

If the right hand oil cooler is damaged,

resulting in oil leakage, the ram air is likely to drive oil up through the ducting to the windshield. Move the regulator control to "OFF" to prevent oil from spattering the inside of the windshield.

(6) FRESH AIR CONTROL.—The fresh air control is located on the central control panel. Turn the regulating knob on the panel counterclockwise to increase the fresh air flow to the cockpit.

(7) SHOULDER HARNESS.

(a) The two free ends of the shoulder harness fit into the safety belt catch and are held securely as long as the catch is closed. To release the harness and safety belt, open the safety belt catch. The belt and harness will immediately fall free.

(b) The harness lock is fastened to the bulkhead behind the seat. The lock has two positions, "LOCK" and "FREE." To release the lock, push the knob of the handle in and lift the handle into the "FREE" position. This allows 18 inches of travel forward with locking provision at intervals of one inch.

WARNING

Under no circumstances should the shoulder harness be omitted, using the seat belt only.

(8) FLYING SUIT RECEPTACLE.—The flying suit receptacle is located on the right hand side of the cockpit, just above the electrical and radio control box.

(9) SEAT ADJUSTMENT. — The pilot's seat is hinged to the armor plate behind the pilot and swings in a vertical arc when the adjusting handle, located on the right hand aft side of the seat, is released. Two springs on the back of the seat tend to pull it up to its most vertical position. A maximum adjustment of four inches is provided.

(10) MAP CASE.—The map case is located on the left hand shelf.

(11) CHART BOARD.—The chart board is stowed in a holder, to the right of the pilot's seat. When in use the chart board is supported by two pegs which fit into adjustable supports on the instrument panel.

(12) REAR VISION MIRRORS.—The three rear vision mirrors are located on the after side of the front sliding section frame.

Hey, Jack. How do ya adjust this seat for optimum comfort?

Section II

NORMAL OPERATING INSTRUCTIONS

1. BEFORE ENTERING THE COCKPIT:

a. NOTE THE FOLLOWING FLIGHT LIMITATIONS.

ITEM	OPERATION OR CONDITION	RESTRICTION
Airplane	Spins	No intentional spinning permitted
Airplane	Inverted flight	10 seconds duration
Airplane	Diving	Dependent on Altitude (see paragraph 19.*d.*)
Wing flaps		
Blow-up Operating (0° to 50°)	Flaps deflected	200 knots
Blow-up Inoperative (0° to 20°)	Flaps deflected	200 knots
50°	Flaps deflected	130 knots
Cabin	Open	300 knots (see paragraph 19.*b.*)
Ailerons	Full throw	300 knots
Cooling Flaps (cowl, intercooler, oil cooler)	Open	No restriction (protected by relief system)
Twin pylon: Drop tank 1000 lb. bomb 1600 lb. bomb	Diving	Normal airplane restrictions apply

THESE LIMITATIONS MAY BE SUPPLEMENTED OR SUPERSEDED BY INSTRUCTIONS INCLUDED IN SERVICE PUBLICATIONS

b. INITIAL GROSS WEIGHT AND LOADING DATA.—Check gross weight and loading with the Handbook of Weight and Balance Data, AN-01-1B-40.

2. ENTRANCE TO CLOSED AIRPLANE.

a. Entrance to the cockpit is gained from the right hand side of the airplane. Three steps and three handgrips are provided. The steps are located in the inboard center section flap, the inboard trailing edge of the center section, and the fuselage. The handgrips are located in the center section just outboard of the step, the instrument access door just below the windshield, and the fuselage.

b. For ease in mounting the airplane, the fuselage step may be opened and pressed down; this action lowers the flaps, thus providing easy access to the center section flap step.

c. To open the cabin, push in the thumb button located on the side panel of the windshield, just forward of the cabin opening, pull out the handle on the side of the sliding section, and push the sliding section aft.

d. Two steps and two handgrips, located on the left hand side of the airplane, permit the mechanic to leave the cockpit at the same time the pilot enters.

3. ON ENTERING THE COCKPIT.

a. BEFORE ALL FLIGHTS:

(1) Adjust the rudder pedals and seat height, if necessary.

(2) Check the controls for freedom of motion and secureness.

(3) Make sure that the mixture control is in "IDLE CUT-OFF," then turn on the battery switch.

(4) Check the fuel and oil quantity aboard.

(5) Make certain that the wings are spread and locked.

(6) Test operate the gun sight illumination. Spare bulbs are located above the instrument panel, left hand side.

(7) See that the gyro horizon and directional gyro are uncaged.

(8) See that the arresting hook control is in the "UP" position. Refer to Section I, paragraph 3.*e.* (3) (*b*).

(9) Check to see that the cabin emergency release pins are secure.

(10) Set the altimeter to the correct barometric pressure.

(11) Test operate the oxygen system as outlined in Section V, paragraph 1.*c.* if the flight planned is an oxygen flight.

(12) Check to ascertain that the desired armament load is carried.

(13) See that all armament switches are in the "OFF" position and that the gun charging valves are in the "SAFE" position.

(14) If rockets are carried, remove the rocket safety plug located on the right hand armament switch box.

(15) Test operate radio and communications controls.

b. BEFORE NIGHT FLIGHTS.

(1) In addition to the standard check in paragraph 3.*a.* above, for night flights turn the battery switch on and check the following items:

(*a*) INTERIOR LIGHTS.

1. COCKPIT LIGHTS. — Check the cockpit lights by turning on the switch on each light, and the rheostat located on the electrical control box.

2. INSTRUMENT BOARD LIGHTS.— Check the instrument board lights by turning on the rheostat, located on the electrical control box. Spare bulbs are carried in a receptacle on the instrument board.

(*b*) EXTERIOR LIGHTS.

1. Check the formation, section, wing, and tail lights by turning on the respective switches and the exterior light master switch. These switches are located on the electrical control box.

4. FUEL SYSTEM MANAGEMENT.

a. GENERAL.—The fuel system is so arranged that drop tank fuel (unprotected) will be used while flying to scenes of combat. Thereafter, for combat and return flight, the fuel from the main tank, or self-sealing drop tank if one is carried, will be used. Fuel from the main tank will be used for starting, warm-up, ground test and take-off.

Figure 23 — Fuel Tank Selector

b. MANAGEMENT.

(1) The fuel system is managed with three controls, viz., the fuel tank selector, the transfer pump switch, and the booster fuel pump switch. The normal flow of gasoline in the system is as follows: the fuel transfer pump transfers gasoline from the selected drop tank to the main tank. From the main tank, the gasoline moves through the booster fuel pump, the fuel selector valve, the fuel strainer, and the engine-driven fuel pump to the carburetor. To effect this flow, the control switches should be set as follows:

(*a*) Fuel selector—"ON."

(*b*) Booster fuel pump switch—"BOOST."

(*c*) Transfer pump switch—"RIGHT" or "LEFT," as desired.

(2) Fuel will flow from the selected drop tank into the main tank until the ball float in the main tank closes the intake valve. The transfer pump will continue working but its pressure is

relieved by a relief valve. The flow will resume when the fuel level drops enough for the valve to open again. This process will continue until the drop tank is empty. At that time an orange warning light on the instrument panel will go on. Turn the transfer switch to the other tank, or, if only one drop tank is being carried, turn the transfer switch to "OFF." The transfer switch shall be in the "OFF" position during take-off and landing.

(3) EMERGENCY BOMB OR DROP TANK RELEASE.—Drop tanks or bombs may be released either manually or electrically. The manual drop tank release should be used only in case of electrical system failure.

(a) To release a drop tank electrically, proceed as follows:

    1. Turn the master armament switch to "BOMBS and GUNS."

    2. Select the tank to be dropped by means of the bomb release switch.

    3. Press the thumb button on the control stick for release.

(b) The manual drop tank release controls are on the left hand shelf (see figure 22). To release a drop tank, select the desired release control and pull it to the limit of its extension.

c. FAILURES.

(1) If there should be a failure in the electrical system while the transfer system is being used, turn the transfer switch "OFF" and turn the fuel selector to the desired tank ("LEFT DROP STANDBY" or "RIGHT DROP STANDBY"). Gasoline will then flow directly from the selected drop tank to the carburetor by action of the engine-driven pump.

Note

When operating with the selector in either of the standby positions, operation will probably be unsatisfactory above 15,000 feet.

(2) In the event that fuel pressure drops, as indicated on the fuel pressure gage, it will be accompanied by a drop in engine performance (unless the fault is in the gage itself). Check the following possible causes:

(a) The selected tank may be empty. Switch to another tank.

(b) The engine-driven fuel pump may fail. In the event of engine-driven pump failure, proceed immediately with the following:

    1. Booster fuel pump—"EMERGENCY."

    2. Fuel selector—"ON."

    3. Shift to neutral blower.

Note

With the booster fuel pump on "EMERGENCY," the fuel pressure will not come up all the way until the supercharger control is shifted to "NEUTRAL."

d. FUEL TANK SELECTION.

(1) The flow of fuel with the fuel selector in its various positions is as follows (see figure 23):

(a) "ON"—from main tank to carburetor.

(b) "RIGHT DROP STANDBY"—from right drop tank to carburetor.

(c) "LEFT DROP STANDBY"—from left drop tank to carburetor.

(d) "OFF"—no flow of fuel.

(2) For all normal activities, including the operation of the fuel transfer system, the selector should be kept in the "ON" position.

(3) Shifts to drop tank positions should be made below 19,000 feet, since it may be impossible to start fuel flow from the selected drop tank above this altitude.

e. FUEL TRANSFER PUMP. — An electric motor driven pump effects the fuel transfer from the drop tanks to the main tank. Solenoid operated valves in the transfer lines provide the means of directing fuel from either of the drop tanks to the main tank. To transfer fuel, set the transfer switch in the desired position ("LEFT" or "RIGHT"). This starts the electric pump and opens the solenoid valve in the line from the selected drop tank to the main tank. The other solenoid valve remains closed, preventing fuel from leaving that tank.

(1) A transfer light, located on the instrument board, will indicate when the selected tank is empty. When the light goes on, turn the transfer switch off, or to the other tank if additional fuel transfer to the main tank is desired.

(2) The pump may cease to transfer fuel above approximately 20,000 feet due to vapor formation. If this occurs, the transfer light will come on even though the drop tank is not empty. It is necessary to descend to a lower altitude before the remaining fuel can be transferred.

f. VAPOR ELIMINATION.—The vapor return line, running from the carburetor to the top of the main fuel tank, returns approximately two quarts of fuel to the main tank in an hour of normal engine operation.

g. BOOSTER FUEL PUMP.—The submerged booster pump is located inside and on the bottom of the main fuel tank. It is controlled by a switch located on the left hand shelf. The switch has three positions, "BOOST," "EMERGENCY" and "OFF." The pump has four specific functions:

(1) It gives a steady flow of vapor-free fuel to the carburetor during high-altitude operation. The pump should be set to the "BOOST" position for all normal high-altitude flight operations. It should be in this position when the engine pump alone will not give the desired fuel pressure.

(2) It serves as an emergency fuel pump in the event of failure of the engine-driven fuel pump. In this event it should be in the "EMERGENCY" position.

(3) It is used in starting. See paragraph 6.

(4) It is used during take-off and landing in the "EMERGENCY" position.

Note

When the booster pump is operated in the "EMERGENCY" position during take-off or landing, or in a ground check with the engine operating, the fuel pressure may rise somewhat.

h. MAIN TANK PRESSURIZING.

(1) The main tank is pressurized by vaporization of the gasoline at high altitudes. This pressure is utilized to prevent excessive fuel losses. An automatic check-relief valve in the vent line closes at altitudes above approximately 18,000 feet in the event that fuel in the main tank is hot enough to boil thereby causing the main tank to pressurize. A pressure relieving action in the vent valve keeps the pressure from exceeding $2\frac{1}{2}$ pounds per square inch.

(2) A manually operated valve is also provided in the vent line for relieving tank pressure. The control for this valve is located on the floor of the cockpit outboard of the right rudder pedal and has two positions, "ON" and "DUMP." It should be left "ON" (aft) at all times except in combat, or as an additional safeguard in the event of a forced landing. In such instances, the control should be pushed forward to the "DUMP" position with the right foot. See figure 2.

i. WATER INJECTION SYSTEM.

(1) OPERATION.—The throttle operates the switch which controls the water injection equipment. To turn the water injection equipment on, lift the lever on the throttle and push the throttle to the most forward position. When the throttle control is in any other position, the water injection equipment is "OFF." The throttle operated switch opens the solenoid shut-off valve on the water regulator and starts the electric water pump. Water pressure acts on a diaphragm to close a jet in the carburetor (deriching the mixture) and to reset the auxiliary stage supercharger regulator, permitting higher carburetor inlet pressure. Water is metered by the water regulator. After metering, the water is mixed with metered fuel just ahead of the fuel spinner nozzle which is located at the face of the main stage supercharger impeller.

(2) WATER SUPPLY.—The water supply is contained in one tank of $13\frac{1}{2}$ U.S. gallon ($11\frac{1}{4}$ Imp. gallon) capacity which is suspended from the bottom engine mount tubes in the accessory compartment. The filler cap can be reached through a door in the left hand upper wing gap cover panel.

(3) FOR COMBAT FLIGHTS.—After engine warm-up, fill the water injection system lines as follows:

(a) With the engine operating at 1200 to 1400 rpm, engage the auxiliary supercharger in "LOW" blower and wait for approximately 30 seconds in order to permit the supercharger to become fully engaged.

(b) Open the throttle to obtain approximately 2200 rpm (all water checks should be made at a minimum of 2200 rpm).

(c) As soon as the engine speed and manifold pressure become stabilized, turn on the water injection limit switch. The switch is readily accessible behind the engine control unit and can be operated with a finger.

(d) The engine will hesitate for approximately one second before the power increases. In addition, proper action of the supercharger reset mechanism will be indicated by a sudden increase in manifold pressure (two to three inches Hg.).

CAUTION

Do not hold the limit switch on any longer than necessary. As soon as a power increase is indicated, turn the switch off.

5. OIL SYSTEM MANAGEMENT.

a. OIL PRESSURES AND TEMPERATURES.—The proper oil pressures and temperatures for the various operating conditions are given on the Power Plant Chart in Section III. The oil temperature can best be kept from exceeding the limit by:

(1) Opening the oil cooler flaps.

Note

The automatic flap control is set so that the flaps will be fully closed at 75°C. (167°F.) and fully open at 95°C. (203°F.).

(2) Reducing engine rpm.

(3) Increasing air speed.

It will be observed that the oil pressure decreases slightly with altitude and takes an additional drop when shifting from "NEUTRAL" to "LOW" or "LOW" to "HIGH." This drop is normal and is to be expected. Oil pressures may drop as low as 70 pounds per square inch at 25,000 feet at rated rpm, full throttle.

6. STARTING ENGINE.

a. PROCEDURE.

(1) Ignition switch—"OFF."

(2) Mixture control—"IDLE CUT-OFF."

(3) Clear engine by pulling propeller through by hand four or five revolutions in normal direction.

WARNING

Never turn over a hot engine by hand.

(4) Fuel selector—"ON."

(5) Fuel transfer switch—"OFF."

(6) Cowl flap switch—"OPEN."

(7) Oil cooler flap switch—"AUTOMATIC."

(8) Intercooler flap switch—"AUTOMATIC."

(9) Propeller control—maximum rpm ("INCREASE").

(10) Supercharger control—"NEUTRAL."

(11) Throttle—Set to red quadrant mark (approximately one inch open).

(12) Battery switch—"ON."

Note

If external power source is available, turn battery switch "OFF" and plug into external power receptacle located beneath the right hand center section.

(13) Booster fuel pump switch — "EMERGENCY."

(14) Primer switch—"ON" 2 to 12 seconds (depending upon temperature and condition of the engine) immediately prior to operating the starter.

(15) Booster Fuel Pump—"OFF."

(16) Turn ignition switch to "BOTH."

(17) Starter switch—"ON."

(18) Move mixture control from "IDLE CUT-OFF" to "AUTO RICH" as soon as engine fires.

Note

Do not pump or move the throttle abruptly until the engine is running smoothly.

CAUTION

If engine should fail to continue running, return mixture control to "IDLE CUT-OFF" IMMEDIATELY to prevent flooding.

(19) Idle at 600 to 800 rpm until normal oil pressure is built up. If oil pressure is not indicated in 30 seconds, stop the engine and investigate.

Note

Normally it should be necessary to operate the starter no more than 30 seconds in order to start the engine. If the starter switch is held "ON" for one minute and the engine does not start, allow the starter to cool before making another attempt.

b. FAILURE TO START ON FIRST ATTEMPT.—If the engine does not start, wait a few minutes to allow any spilled fuel to drain out of the blower drain. Inspection of the exhaust pipe outlets, especially those from the upper cylinders, should indicate whether the engine has been over or under-primed. No trace of smoke indicates under-priming; excessive black smoke indicates over-priming. The use of the primer switch should be governed accordingly. If the engine is over-primed, clear the cylinders and induction system of the excess fuel as follows:

(1) Mixture control—"IDLE CUT-OFF."
(2) Booster fuel pump switch—"OFF."
(3) Ignition switch—"OFF."
(4) Battery switch—"OFF."
(5) Throttle—full "OPEN."
(6) Clear engine by turning propeller four or five revolutions by hand in normal direction.

WARNING

Never turn over a hot engine by hand.

7. WARM-UP AND GROUND TEST.

a. GENERAL.—For warm-up and ground testing, the following should be observed:

(1) Propeller control—full "INCREASE."
(2) Cowl flaps—"OPEN."
(3) Oil cooler flaps—"AUTOMATIC."
(4) Intercooler flaps—"AUTOMATIC."
(5) Mixture control—"AUTO RICH."
(6) Cylinder head temperature — 232°C. (450°F.) maximum. If cylinder head temperatures approach 232°C. (450°F.), the engine should be cooled at 1000 rpm before continuing with the ground test.

b. ENGINE WARM-UP.

(1) Check oil pressure. With cold oil, pressure may be above 200 pounds per square inch until oil-in temperature is approximately 40°C. (104°F.)

(2) Idle at 1000 rpm until oil temperature is 40°C. (104°F.).

(3) Ignition Safety Check.

Engine speed approximately 1000 rpm. May be performed during warm-up.

(*a*) Ignition switch from "BOTH" to "RIGHT" and back to "BOTH."
(*b*) Ignition switch from "BOTH" to "LEFT" and back to "BOTH."
(*c*) Ignition switch to "OFF" (momentarily) and back to "BOTH."

Slight drop-off of rpm on each separate magneto, and complete cutting out of engine at "OFF" position indicates proper connection of ignition leads.

c. ENGINE AND ACCESSORIES GROUND TEST.

(1) Open throttle briefly to at least 2200 rpm or 30 inches Hg. and check the following:

CAUTION

Backfiring may result from opening the throttle too suddenly from the idling position in flight or on the ground.

(*a*) Oil pressure 90 to 95 pounds per square inch.

(*b*) Fuel pressure 16 to 18 pounds per square inch.

(2) MAGNETO CHECK.

(*a*) Adjust throttle to give an engine speed of 2200 rpm.

(*b*) Rpm drop-off should be no more than 100 when shifting from "BOTH" to "RIGHT" or "LEFT."

(*c*) Permit operation to stabilize on "BOTH" after operation on one set of plugs before checking the other set.

(3) PROPELLER GOVERNOR CHECK.

(*a*) Adjust the throttle to give an engine speed of 2000 rpm.

(*b*) Move the propeller control from full "INCREASE" to "DECREASE." The engine speed should drop to about 1200 rpm.

(*c*) Return the propeller control to "INCREASE."

(4) IDLE MIXTURE CHECK.

(*a*) Set throttle for 600 rpm.

(*b*) Booster fuel pump—"OFF."

(*c*) Move the mixture control lever smoothly and steadily into "IDLE CUT-OFF" and observe the tachometer for any change in rpm.

(*d*) Return the mixture control to "AUTO RICH" before the engine cuts out. A rise of more than 10 rpm indicates too rich an idle mixture, and no change or drop in rpm indicates that the mixture is too lean. A rise of 5 to 10 rpm is recommended in order to permit idling at low speeds without danger of fouling plugs and, at the same time, to afford good acceleration characteristics.

CAUTION

During idling operations, do not operate the booster fuel pump in "EMERGENCY" position, as excessive fuel pressures may result which would cause high fuel flows. This would tend to foul the plugs.

(5) SUPERCHARGER CHECK AND DE-SLUDGING PROCEDURE.—The supercharger check should never be made nor the couplings desludged until the oil temperature has reached 40°C (140°F), and it is preferable to wait until the oil

temperature has reached 60°C (140°F). If there is not enough time to complete a regular supercharger check, desludge the couplings twice as directed in paragraph (f) below.

CAUTION

When shifting from one blower ratio to another, be sure to shift quickly, without dwelling between positions. This is to avoid dragging and slipping the couplings.

(a) Adjust the throttle to obtain 1400 rpm; then shift the supercharger control from "NEUTRAL" to "LOW."

(b) Wait at least 30 seconds; then shift from "LOW" to "HIGH."

(c) After a minimum of 30 seconds in "HIGH", open the throttle to obtain 30 inches of manifold pressure, and note the rpm.

(d) Shift back to the "LOW" position, and after the manifold pressure has stabilized, readjust the throttle to obtain 30 inches of manifold pressure. Note the rpm.

(e) Shift from "LOW" to "NEUTRAL", and again adjust the throttle to obtain 30 inches of manifold pressure. Note the rpm. An increase in engine rpm when shifting to a lower blower ratio while maintaining a constant manifold pressure indicates that the couplings are operating correctly.

(f) Desludge the couplings by shifting as directed in paragraphs (a), (b), and (c) above. After a minimum of 30 seconds operation in high blower, move the supercharger control directly back to "NEUTRAL."

(6) ELECTRICAL CHECK WITH ENGINE RUNNING.

(a) Disconnect external power source, if used.

(b) Make sure the battery switch is on.

(c) Turn on some electrical equipment such as the cockpit or instrument lights.

(d) Run the engine rpm up past 1400 (to close the reverse current cut-out).

(e) Turn off the battery switch. If the lights turned on in step (c) stay on, the reverse current cut-out has closed.

(f) Increase the engine rpm and watch the voltmeter. The voltage should increase to about 28 volts and stay there regardless of any further increase in engine rpm.

(g) If the reverse current cut-out does not close, or if the voltmeter reading does not lie beeween 27.5 and 28.5 volts, corrective steps should be taken before take-off.

(7) HYDRAULIC PRESSURE CHECK.—Check the hydraulic pressure gage. It should indicate 900 to 1150 pounds per square inch.

(8) RADIO CHECK.—Test the radio operation (refer to Section V, paragraph 2).

(9) AUTOMATIC SPARK ADVANCE CHECK.

(a) Propeller control — maximum rpm ("INCREASE"). Prop full low pitch.

(b) Mixture control—"AUTO RICH."

(c) Open throttle to 1700 to 1800 rpm.

(d) Check rpm drop-off when operating on one magneto. If engine speed remains constant, the spark has advanced and is, therefore, in proper adjustment. If engine drops off the same as when running a normal magneto check, the spark has probably not advanced and is out of adjustment.

(10) CHECK OPERATION OF THE WING FLAPS.

Last Guy Off the Field's a Rotten Egg!

8. SCRAMBLE TAKE-OFF.

It is possible to make an emergency take-off providing the oil temperature is above 40°C.

(104°F.) prior to take-off. In cases of extreme emergency, where the above temperature cannot be met, run the engine up; if it does not operate roughly or cut out altogether, take off.

9. TAXIING INSTRUCTIONS.

a. Use the S-turn procedure for adequate forward vision on taxi strips. However, let the airplane roll freely where possible, using the brakes as an aid in steering, stopping, and holding only.

b. Use the tail wheel lock in extended crosswind taxiing to relieve excessive braking action.

c. Use low power when taxiing. Don't rev up the engine and then ride the brakes. Bear in mind that badly over-heated brakes are not fully effective and can fuse the brake discs to the extent of leaving them frozen for landing.

d. Keep electrical load at a minimum to prevent battery discharge.

10. TAKE-OFF.

a. Refer to the Take-Off, Climb and Landing Chart in Appendix I.

b. CHECK LIST.

(1) Shoulder harness and safety belt—secure and locked.

(2) Fuel tank selector—"ON."

(3) Mixture control—"AUTO RICH."

(4) Booster fuel pump—"EMERGENCY."

(5) Supercharger control—"NEUTRAL."

(6) Propeller control—maximum rpm ("INCREASE").

(7) Cowl flaps—"OPEN."

(8) Intercooler flap switch—"AUTOMATIC."

(9) Oil cooler flap switch—"AUTOMATIC."

(10) Rudder tab control—six degrees "NOSE RIGHT."

(11) Aileron tab control — four degrees "RIGHT WING DOWN."

(12) Elevator tab control — one degree "NOSE UP."

(13) Wing flaps — set as required. Refer to paragraph 10.c.(1) below.

(14) Tail wheel—"LOCKED."

(15) Check magnetos and cylinder head and oil temperatures.

c. GENERAL.

(1) FLAP SETTINGS.—For normal operation it is recommended that a flap setting of 20 degrees be used for take-off. Actually, any flap setting from zero degrees to 50 degrees (full down) may be used, the higher settings giving shorter ground distance. Take-offs with flaps up are easily accomplished with a small increase in run, dispensing with the inconvenience of retracting the flaps after take-off. In addition, the rate of climb immediately after take-off with flaps deflected is inferior to that with flaps "UP." Take-off at high flap settings and at full flap should be made only when it is necessary to obtain the shortest possible deck run and after more experience with settings increased gradually from the recommended setting of 20 degrees. When a high flap setting is used, the elevator tab should be set slightly more tail heavy (about one degree).

Note

It has been found that with the flaps down the tail cannot be held on the ground, with the stick full back, at manifold pressures greater than 44 inches Hg. Also, when operating from a wooden platform the wheels will start slipping on the deck at approximately the same manifold pressure. As a result, when making a carrier take-off it is necessary to advance the throttle through the final portion of its travel as the airplane starts to roll. No difficulty should be encountered in this operation.

(2) TAB SETTINGS.—Due to the high engine power and low propeller reduction ratio, the proper tab settings must be used for take-off; otherwise, needlessly high forces will be encountered. The rudder force required to maintain a straight run will be high unless the rudder tab has been set at approximately six degrees "NOSE RIGHT" prior to the take-off run. Also, the left wing tends to be slightly heavy just as the airplane becomes airborne, due to high torque reaction. If the aileron tab is set approximately four degrees "RIGHT WING DOWN" before the start and if the airplane is not lifted off prematurely, this effect can be avoided. Use of the proper tab settings is particularly important when high flap settings and maximum power are used. Individual airplanes will require slightly different tab settings from those given above. It may be noticed that the tab control knobs rotate slightly when the stick and rudder controls are moved. However, the actual tab setting does not change.

d. MINIMUM RUN TAKE-OFF.

(1) Wing flaps—full "DOWN" (50 degrees).

(2) Propeller governor—full "INCREASE."

(3) Elevator tabs — Three to four degrees "NOSE UP."

(4) Hold brakes slightly until tail starts to rise.

(5) Release brakes and allow tail to rise to near level flight attitude (tail high).

(6) Take off when minimum flying speed is attained (approximately 70 knots indicated). The nose will be slightly heavy. If the take-off is made from an unpaved or muddy runway, take off with the tail slightly lower than directed above.

Note

If an obstacle is to be cleared during take-off, the wing flap setting should be reduced to approximately 30 degrees.

e. CATAPULT TAKE-OFF.

(1) Shoulder harness and safety belt—tight.

(2) Tighten engine control friction adjustment knob.

(3) Place back and head firmly against seat and head rest.

(4) Place feet against rudder pedals with legs stiff.

(5) Brace right arm.

(6) Push throttle full forward and grasp catapult throttle hold.

11. ENGINE FAILURE DURING TAKE-OFF.

a. In the event of engine failure during take-off, LAND STRAIGHT AHEAD.

b. As many as possible of the operations listed below should be performed in the order given:

(1) Release drop tank or bombs.

(2) Landing gear — "UP" unless sufficient runway is available STRAIGHT AHEAD for a landing in the normal (wheels down) landing condition.

(3) Wing flaps—full "DOWN."

(4) Lower the seat several notches.

(5) Switches (battery and ignition)—"OFF."

(6) Fuel selector—"OFF."

The battery switch turns off the booster fuel pump and the fuel transfer pump.

12. AFTER TAKE-OFF.

a. FOR MOST EFFICIENT OPERATION:

(1) Reduce manifold pressure to not over 43.5 inches Hg.

(2) Reduce rpm to not over 2600.

(3) Retract landing gear.

(4) Retract wing flaps.

(5) Trim airplane for 130 knots indicated air speed for best climb.

(6) Adjust cowl flaps so as not to exceed maximum cylinder head temperature.

(7) Set transfer switch to desired tank, and turn booster pump switch to "OFF."

(8) Unless a rapid rate of climb is desired, it is recommended that the manifold pressure and rpm be further reduced to 34.5 inches Hg. and 2400 rpm, and the air speed increased to 10 knots above the normal air speed for climb.

(9) Mixture control — "AUTO LEAN."

CAUTION

Do not retract the flaps too soon or too rapidly after take-off if the speed is very low; otherwise, the airplane may settle due to the loss in lift. It should be remembered that the higher the take-off speed, the better the control.

13. CLIMB AND LEVEL FLIGHT.

a. MILITARY POWER (5 MINUTES, 2800 RPM).—Operate in accordance with the Power Plant Chart in Section III and the Engine Calibration Curve in Appendix I. Inasmuch as the cylinder base temperatures might exceed the high limit of 177°C. (350°F.) when the cylinder head temperatures exceed 245°C. (473°F.), the maximum allowable cylinder head temperature has been reduced from 260°C. (500°F.) to 245°C. (473°F.)

b. RATED POWER (MAXIMUM CONTINUOUS, 2600 RPM WITH CYLINDER HEAD TEMPERATURE 232°C. (450°F.) OR UNDER).—Operate in accordance with the Power Plant Chart in Section III and the Engine Calibration Curve in Appendix I.

CAUTION

DO NOT OPERATE WITH CYLINDER HEAD TEMPERATURES IN EXCESS OF 232°C. (450°F.) FOR MORE THAN 60 MINUTES. DO NOT EXCEED 245°C. (473°F.) CYLINDER HEAD TEMPERATURE AT ANY TIME.

14. GENERAL FLYING CHARACTERISTICS.

a. Refer to the Flight Operation Instruction Charts in Appendix I for the effects of changes in gross weight or any external resistance and to the Power Plant Chart in Section III for engine operating data.

b. STABILITY.—The longitudinal stability becomes quite low when the center of gravity approaches the aftermost limit of the permissible center of gravity range in low-speed, high-power flight conditions such as climb and carrier approach. In the cruising condition the airplane has a high degree of stability at all permissible center of gravity positions.

c. TRIM CHANGES.

(1) Extension of the landing gear and the wing and cooling flaps changes the trim of the airplane only slightly. However, for the convenience of the pilot, the direction of the trim changes is listed as follows:

(a) Extend landing gear: — tail-heavy as gear begins to extend, but nose-heavy when the gear is fully down.

(b) Extend wing flaps:—tail-heavy at small flap angles, and nose-heavy at large flap angles.

(c) Open cowl flaps:—nose-heavy.

(d) Open oil cooler flaps:—tail-heavy.

(e) Open intercooler flap:—tail-heavy.

(2) The airplane exhibits no unusual flying characteristics at low speeds. There is some change of both lateral and directional trim due to the application of power at low air speeds. However, the effectiveness of the aileron and rudder trim tabs is sufficient to offset these changes of trim easily. All of the trim tabs on this airplane are effective and sensitive.

d. COMBAT POWER — 2800 RPM (FIVE MINUTES).

(1) Combat ratings are based upon engine structural limitations, water injection being used to suppress detonation. Combat Power is used primarily for combat.

CAUTION

Because Combat Power places a strain on the engine, it is to be used with discretion and should be treated as combat ammunition which is expended unhesitatingly but only when the occasion demands.

(2) To obtain Combat Power:

(a) Mixture control—"AUTO RICH."

(b) Propeller control — maximum rpm ("INCREASE").

(c) Throttle—lift finger lever and move to full forward position.

WARNING

Do not use Combat Power when the fuel supply is nearly exhausted in the tank which is being used. If the fuel supply in the tank were completely exhausted, the sudden surge of water to the engine would stop it.

(3) When the water supply is exhausted while operating at Combat Power in high blower at or below critical altitude, watch the carburetor air temperature warning light. If the light is on, follow the corrective procedure outlined in Section I, paragraph 3.b.(9).

(4) When in low or high blower, the supercharger regulator resets to normal when the water supply is exhausted. WHEN IN NEUTRAL BLOWER BELOW 3000 FEET, THE THROTTLE MUST BE RETARDED TO PREVENT OVERBOOSTING WHEN THE WATER SUPPLY IS EXHAUSTED.

e. CRUISING.—The engine should be operated in "AUTO LEAN" for cruising power operation. A cylinder head temperature limit of 232°C. (450° F). is not to be exceeded for continuous cruising.

(1) MAXIMUM.—While cruising operations may be conducted at any engine power below normal rated power, if minimum fuel consumption is of importance and if it is tactically feasible to do so, cruising operation should be conducted at power considerably below maximum cruising. For specific details refer to the Flight Operation Instruction Charts in Appendix I.

(2) RECOMMENDED.

(a) Sea level operation at approximately 155 knots indicated air speed at 1300 rpm, neutral blower, will result in near best range operation (approximately 38 U. S. gallons, 32 Imp. gallons, per hour).

(b) When cruising in low or high blower at air speeds less than 190 knots, a reduction in drag may be obtained by manually closing the intercooler flaps. If an rpm switch is incorporated in the airplane, it will not be necessary to close the intercooler flap manually during cruising operations in low or high blower. The intercooler flap switch should be returned to the "AUTOMATIC" position before entering maneuvers requiring the use of high power.

f. SUPERCHARGER OPERATION. — In general, confine shifting to engine speeds between 1400 rpm and 2600 rpm. However, in an emergency, a shift can be made at Combat and at Military Power. Do not use "LOW" blower when the de-

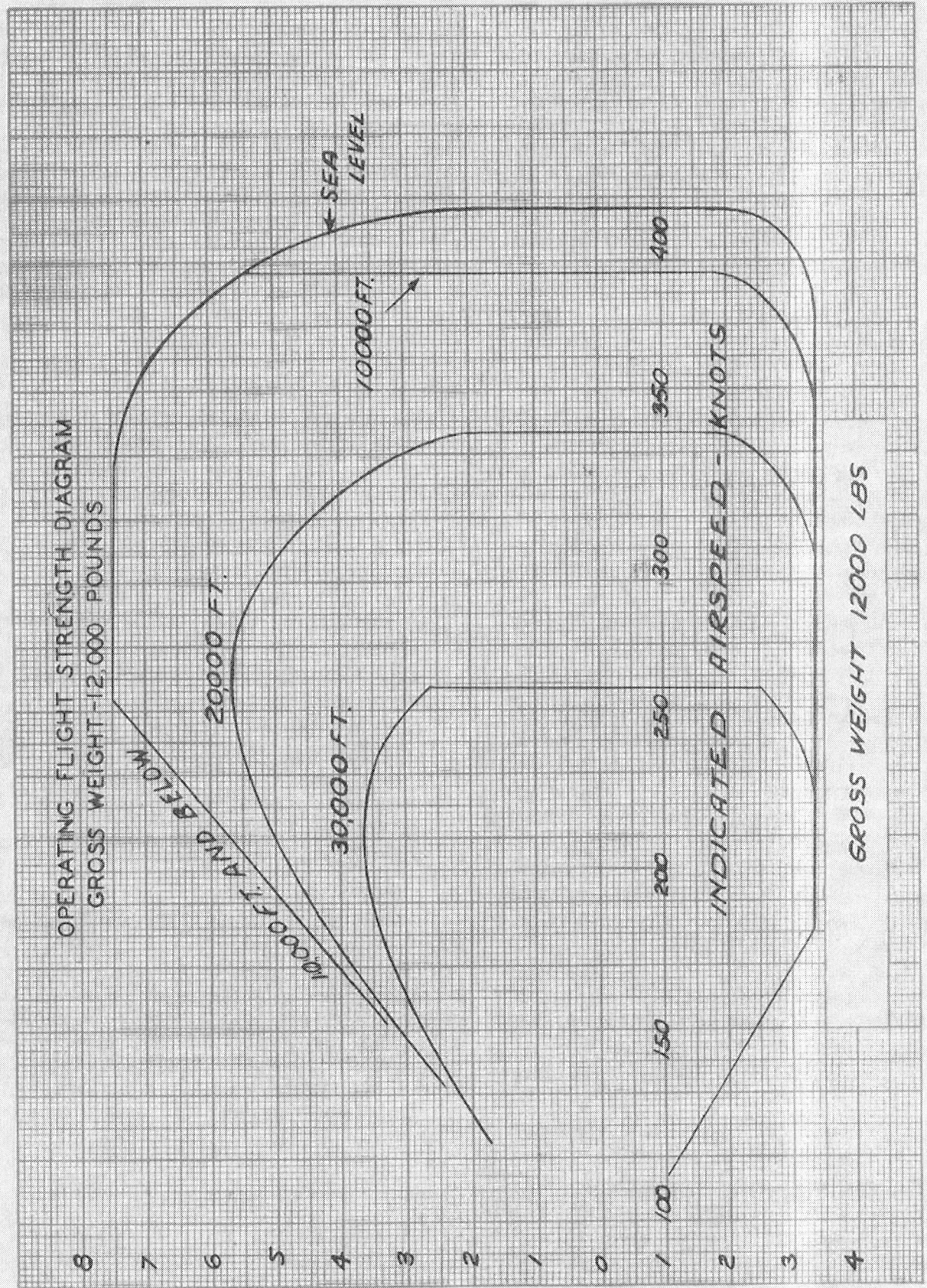

Operating Flight Strength Diagram

Section II
Paragraphs 14-15

sired power can be obtained in "NEUTRAL," or "HIGH" when the desired power can be obtained in "LOW." Otherwise, high carburetor air temperature and uneconomical fuel consumption will result. When necessary, the supercharger control may be moved directly from "NEUTRAL" to "HIGH" and vice versa without stopping at the "LOW" position. When shifting to a lower blower ratio, it is not necessary to change either the throttle position or the engine speed. The shift is smooth with an almost imperceptible surge of power.

(1) At engine speeds above 1900 rpm or manifold pressure above 30 inches Hg., the shift to a higher blower ratio should be made as follows:

(a) Mixture control—"AUTO RICH."

(b) Move the supercharger control to the desired blower ratio.

(c) As soon as the manifold pressure begins to increase, quickly move the throttle to the 2/3 open position.

(d) When the manifold pressure has stabilized, readjust the throttle to obtain the desired manifold pressure.

(2) At engine speeds above 1900 rpm or manifold pressures above 30 inches Hg., the throttle should be retarded when shifting in order to prevent manifold pressure surges above the allowable manifold pressures shown on the engine calibration curve for low and high blower. At speeds and manifold pressures lower than 1900 rpm or 30 inches Hg., the throttle need not be retarded since the manifold pressure surges will be of much smaller magnitudes.

g. VIBRATION.—Operation is smooth except in a range between 1850 and 2050 rpm. Operation in this range is not dangerous but may be objectionable and if so, should be avoided.

h. SURGING.—Avoid part throttle operations at high rpm with low manifold pressures in either low or high blower: power pulsations and surging are likely to result. If surging is encountered:

(1) Increase manifold pressure (open throttle).

(2) Reduce engine rpm or shift to a lower blower and adjust power to the value desired.

i. MAXIMUM PERMISSIBLE INDICATED AIR SPEEDS AND ACCELERATIONS.—The maximum permissible speeds and accelerations at various altitudes are shown on the graph above for a gross weight of 12,000 pounds. At other weights the permissible accelerations are such as to maintain a constant product of gross weight and acceleration except that 7.5g positive and 3.4g negative should not be exceeded. The limit of the actual accelerations and speeds that can be withstood with safety is indicated by a general buffeting or shaking. It is dangerous to continue increasing acceleration or speed once the buffet begins because the shaking and vibration increase the loads in the tail structure and so may cause damage to the stabilizer and elevator; therefore, when buffeting is encountered, immediately reduce speed or acceleration or both. In general, this phenomenon occurs at lower indicated speeds at the higher altitudes as seen on the graph below.

WARNING

TESTS TO DATE INDICATE THAT THE AIRPLANE, IF PERMITTED, WILL ATTAIN MUCH HIGHER THAN THE LIMIT SPEEDS. PILOTS SHOULD GAIN FAMILIARITY WITH THE DIVING CHARACTERISTICS OF THE AIRPLANE GRADUALLY, WHILE MAINTAINING ABSOLUTE CONTROL OVER THE DIVING SPEED. THE SPEED LIMITS SHOULD BE FIRST APPROACHED ONLY AFTER SUCH FAMILIARIZATION, AND IN DIVES AT MODERATE ANGLES. FAMILIARITY WITH THE LIMITATIONS IMPOSED BY THE SHAKE DISTURBANCE ON PULL-OUT SHOULD ALSO BE GRADUALLY ACQUIRED.

15. MANEUVERS.

a. MANEUVER FLAPS.—The wing flaps have been designed for possible use in maneuvering. The flaps may be used to increase the lift and thereby decrease the radius of turns at low speeds. The flaps are also useful in increasing the drag of the airplane so that it may be quickly decelerated to the optimum speed for a short radius turn. In general, flap deflections of 20 degrees or less will be most helpful in improving maneuverability. Therefore, a setting of 20 degrees has been established as the "maneuver flap" condition.

CAUTION

Flaps are not to be used at speeds greater than 200 knots. As stated above, the flaps have also been designed for use in maneuvering the airplane in combat. With typical maneuvering flap deflections of 20° or less, the airplane may be maneuvered at equivalent "flaps up" accelerations up to 200 knots.

b. AILERONS.—The use of the ailerons is subject to the following restrictions:

Amount of Lateral Stick Throw	Maximum Speed
Full	300 knots

16. STALLS.

a. GENERAL.

(1) The stalling characteristics of the airplane are not abnormal, and warning of the approach of the stall exists in slight tail buffeting, the abnormal nose-up attitude, and increasing left wing heaviness with power on. The center of gravity positions enforced by recent combat requirements are farther to the rear than would be normally desirable. This results in a low degree of longitudinal stability under high power conditions at low air speeds. While the elevator forces are generally normal in direction, they vary only a small amount in approaching the stall with power on, and the control movement is very small. Thus the elevator control force and position do not provide the normal degree of "feel" or warning of change in air speed or angle of attack. Pilots should observe carefully and familiarize themselves with this characteristic in the landing approach condition and in maneuvering turns which approach the stall at higher speeds. This should be done at various flap positions and powers until pilots are thoroughly familiar with the airplane in these conditions.

(2) The stall with power on is rather pronounced, particularly with flaps down, but is preceded by some warning in the nature of buffeting. In the carrier approach condition, the approach to the stall is indicated to some extent by increasing left wing heaviness and the increasing amount of right rudder required. The stall in this condition (flaps "DOWN," power "ON") is accompanied by a relatively sharp roll to the left.

Note

Pilots should familiarize themselves thoroughly with the stall, in both straight flight and tight turns.

(3) The indicated stalling speeds for an 11,300 pound fighter are given in the table following.

CONDITION	FLAPS	POWER	INDICATED STALLING SPEED KNOTS
Landing	50°	Closed Throttle	75
Landing	30°	Closed Throttle	77
Landing	20°	Closed Throttle	79
Landing	50°	Power on (Level Flight) 23 In. Hg., 2400 Rpm.	66
Clean	Up	Closed Throttle	87
Clean	Up	Power on (Level Flight) 18 In. Hg., 2400 Rpm.	84

17. SPINS.

WARNING

NO INTENTIONAL SPINNING OF THE MODEL F4U-4 AIRPLANE IS PERMITTED.

a. GENERAL.—If a spin should inadvertently develop in either the clean or landing condition, the pilot should apply full opposite controls, viz., controls against the stops. In recent spin tests, little difficulty was experienced in recovering from spins of four turns in each direction in the clean condition and from one turn in the landing condition. Use of ailerons against the spin will improve the recovery characteristics over ailerons neutral. The average recovery turns for a four-turn spin in the clean condition are about 1½ turns. In the landing condition the average recovery from a one-turn spin is ¼ turn.

b. RECOVERY TECHNIQUE.—As a result of spin tests, it is recommended that if a normal spin is inadvertently entered the following steps for recovery be initiated immediately:

(1) Apply full opposite controls *sharply*, leading with opposite rudder, and follow by applying full forward stick. Apply ailerons *against* the spin.

(2) Hold *full* reversed controls until rotation stops and airplane assumes normal diving attitude.

(3) Ease airplane out of ensuing dive. Do not pull stick back too rapidly as a high speed stall may result requiring more altitude for recovery.

(4) The rate of rotation will probably increase after full opposite controls are used. Don't be alarmed; this is a good sign and recovery is starting.

(5) Use tabs if forces are too heavy, especially the elevator tab. The latter is effective in reducing push forces during spin recovery.

(6) Oscillation is present in left spins. The nose oscillates between a position varying from approximately on the horizon to 40 degrees to 50 degrees below the horizon. This does not mean

that a flat spin is developing. Recovery will be normal. Recovery will be faster if controls are reversed when the nose is at the steeper angle in the oscillation.

(7) If full opposite controls cannot be held and the stick walks back, return the controls with the spin for a brief interval, and repeat full recovery control.

Note

Full forward stick (stick against stop) must be applied for spin recovery in this airplane. Make certain that full reversed controls are held until recovery is effected.

18. PERMISSIBLE ACROBATICS.

a. All normal acrobatics are permissible when not carrying bombs, tanks, or similar loads, if the following precautionary measures are observed:

(1) Do not exceed allowable speeds and accelerations.

(2) Inexperienced pilots shall not enter loops or Immelmans at less than 280 knots indicated air speed. This speed may be lessened slightly as more experience is gained in these maneuvers.

(3) Inexperienced pilots shall not enter slow rolls at less than 180 knots indicated air speed.

(4) Inverted flight shall not exceed 10 seconds duration because of loss of oil pressure and possible damage to bearings, especially the thrust bearing, caused by the temporary insufficient engine lubrication.

19. DIVING.

a. CHECK LIST.

(1) Windshield defroster—"ON."
(2) Cabin—closed.
(3) Landing gear control—"UP."
(4) Dive brake control—as desired.
(5) Wing flap control—"UP."
(6) Propeller control—Set as desired between 2050 and 2250 rpm.
(7) Mixture control—"AUTO RICH."
(8) Throttle—Slightly "OPEN."

Note

Fifteen to twenty inches of manifold pressure is recommended during prolonged dives. Manifold pressures much below fifteen inches, if held in a prolonged dive, will foul up the engine in the same manner as do prolonged glides with closed throttle. Caution should be observed in diving from a high altitude, as manifold pressure will build up rapidly at a constant throttle setting. Caution should also be observed to open the throttle slowly at completion of dive so partly cooled engine will not cut out.

(9) Supercharger—"NEUTRAL."

Note

Neutral blower should be used for all dives except those incident to military tactics at high altitudes.

(10) Fuel Tank Selector—"ON."
(11) Cowl Flaps—"CLOSED."
(12) Oil Cooler Flaps—"CLOSED."
(13) Intercooler Flaps—"CLOSED."
(14) MAXIMUM RPM LIMIT 3120 RPM (NOT OVER 30 SECONDS DURATION.)

b. COCKPIT CABIN.—The cockpit cabin sliding section must be closed before entering high speed dives as it is not designed for such speeds in the open position. In the open position speeds up to 300 knots are allowable.

c. DIVE BRAKE CONTROL. — When the dive brake control is operated at high indicated air speeds, the wheels will trail instead of extending fully and locking but are nevertheless effective as a dive brake.

20. APPROACH AND LANDING.

a. CHECK LIST.

(1) Shoulder harness—locked.
(2) Tail wheel — "LOCKED" (for field); "UNLOCKED" (for carrier).
(3) Fuel tank selector—"ON."
(4) Fuel transfer pump—"OFF."
(5) Booster fuel pump—"EMERGENCY."
(6) Mixture—"AUTO RICH."
(7) Supercharger control—"NEUTRAL."
(8) Propeller control—2400 rpm.
(9) Cowl flap switch—"CLOSED."
(10) Landing gear — "DOWN".

(11) Wing flaps—set 50 degrees, or as required, for field landing, 50 degrees for carrier,

(12) Arresting hook — "UP" for field; "DOWN" for carrier.

(13) Master Armament Switch—"OFF."

(14) Gun charging knobs—"SAFE" (push in).

(15) If rockets are carried, remove the rocket safety plug located on the right hand armament switch box.

(16) Tabs—Set as required for landing condition.

b. RECOMMENDED SEQUENCE.

(1) Observe items on check list.

(2) Open cabin.

(3) Air speed in approach — approximately 95 knots.

c. FLAP SETTING.—For field landing, it is recommended that a setting of 50 degrees be used. Lesser flap settings will result in increased ground run. Flaps full down (50 degrees) shall be used for all carrier landings.

d. CROSS WIND LANDING.— Cross wind landings can best be made by landing with tail slightly up and somewhat less than normal flap angle (about 30 degrees), all other normal landing conditions being about the same. Use some downwind rudder just prior to contact with the ground to head the airplane in the direction of motion over the ground. During the run after landing, there will be a tendency for the up-wind wing to rise, and the airplane will turn into the wind. Use a little rudder or brake for counteraction.

WARNING

Use the brakes cautiously until the tail wheel is on the ground.

e. MINIMUM RUN LANDING.

(1) Flap setting — full "DOWN" (50 degrees).

(2) Propeller governor — maximum rpm ("INCREASE").

(3) Throttle—slightly "OPEN."

(4) Indicated air speed — Approximately 95 knots.

(5) The approach should be rather flat as in a carrier landing; the nose should be high. Bring the airplane in about ten feet above the runway, close the throttle and drop the airplane to the runway. Use the brakes as necessary.

f. TAKE-OFF IF LANDING IS NOT COMPLETED.

(1) In the event that landing is not completed, open the throttle smoothly. Retract the landing gear immediately and open the cowl flaps. The oil cooler flaps and intercooler flaps will operate automatically. When the air speed has reached 110 knots, raise the flaps. This procedure should be followed to avoid overheating the power plant, which would result if extended operations were undertaken in the landing condition. It will probably be advisable to re-trim the airplane after advancing the throttle. This is done in order to relieve the slightly increased forces on the surface controls.

21. STOPPING OF ENGINE.

a. Desludge the hydraulic couplings in accordance with paragraph 7.c.(5) of this section.

b. Before shutting down the engine, use the following procedure for cooling:

(1) Cowl flaps—full "OPEN" while idling and for at least 10 minutes after stopping.

(2) Intercooler flap—"OPEN" to cool accessory compartment.

(3) Oil cooler flaps—"OPEN."

(4) Propeller control—maximum rpm ("INCREASE").

(5) Throttle—Set for 800 to 1000 rpm (near the red quadrant mark) to cool cylinder head temperatures to approximately 200°C. (392°F.) or below.

c. TO STOP THE ENGINE:

(1) Booster fuel pump—"OFF."

(2) Mixture control—"IDLE CUT-OFF."

(3) Ignition switch—"OFF" when the propeller stops turning.

Note

As soon as the engine begins to cut out, move the throttle forward slowly to fill the carburetor with fuel for the next flight.

(4) Battery switch—"OFF."

(5) Fuel selector—"OFF."

(6) Turn off all switches used for flight (radio, lights, etc.).

Note

Cowl flaps, oil cooler flaps, intercooler flap and cockpit cabin must be closed as soon as engine is cool.

d. OIL DILUTION PROCEDURE.

Note

THE OIL DILUTION SYSTEM IS NOT INSTALLED IN THE AIRPLANE AT

DELIVERY. ONE OIL DILUTION KIT, CONTAINING THE COMPLETE SYSTEM, IS FURNISHED WITH EVERY TENTH AIRPLANE FOR INSTALLATION IN EXTREME COLD WEATHER CONDITIONS.

(1) In the event of a low temperature forecast, viz., below —5°C. (+23°F.), the oil in the warm-up circuit shall be diluted in the following manner:

(a) Open the manual shut-off valve in the oil dilution line. This valve is located on the side of the left hand oil tank support in the engine accessory compartment.

(b) Start engine (see paragraph 6, this section).

(c) Engine speed constant — 1000 rpm.

(d) Oil dilution switch — "ON" (APPROXIMATELY FOUR MINUTES).

(e) Stop engine by moving mixture control to "IDLE CUT-OFF."

(f) Hold oil dilution switch on until engine stops.

(g) When a cold engine in which the oil was diluted prior to shut-down is subsequently started and, after running a short while, the oil pressure starts to fluctuate or drop, the dilution valve shall be opened intermittently for intervals of a few seconds over a period of about 15 seconds. If the oil pressure still does not steady out, stop the engine and wait for approximately five minutes before attempting another start.

(2) PRECAUTIONS.

(a) Do not overdilute.

(b) Guard against fire.

(c) Dilute only when justified by a forecast of low temperatures, viz., below —5°C. (+23° F.)

(d) Allow adequate warm-up before taking off, except in cases of extreme emergency.

(e) Keep the oil system free from sludge and water.

(f) Check position of dilution line shut-off valve.

(g) Since the oil in the hydromatic propeller is not diluted, care must be taken to determine that the propeller pitch-changing mechanism is operating prior to take-off.

22. BEFORE LEAVING COCKPIT.

a. Install the surface control lock. There are two fittings on the floor adjacent to the center control panel. Pull up the plungers in these fittings and insert the V-shaped tube assembly in place in the fittings. Secure the clamp around the control stick and tighten the pivoted legs against the left and right brake pedals. The knurled nuts on the legs provide a means of adjusting the legs against the pedals and tightening.

CAUTION

Make sure that the surface controls are locked whenever the airplane is parked; otherwise, damage to the rudder, elevators or ailerons may result.

23. MOORING.

a. In the event of unfavorable weather conditions involving high wind velocities or heavy precipitation, or if the airplane is to remain parked for a considerable length of time, make certain that it is tied down securely. The following procedure shall be used:

(1) Lock the tail wheel.

(2) Place chocks fore and aft of the front wheels.

(3) Secure the tail by means of the hold back link on the tail wheel housing.

(4) Keep the wings spread. Secure the wings by means of the tie-down links in the outer panels.

(5) The landing gear drag links and towing links may also be used for tieing down.

Section III
OPERATING DATA

1. AIR SPEED CORRECTION TABLE.

The calibration below represents the air speed head (pitot tube) position error and gives the corrected indicated air speed for a given reading of the cockpit air speed indicator assuming zero scale error for the instrument itself.

Cockpit Air Speed Indicator Reading in knots	Clean Condition — Flaps Up		Landing Condition — Flaps Down	
	Correct Indicated Air Speed in knots	Correction in knots	Correct Indicated Air Speed in knots	Correction in knots
70	—	—	67	−3
80	—	—	78	−2
90	88	−2	89	−1
100	99	−1	100	0
110	110	0	111	+1
120	122	+2		
130	132	+2		
140	143	+3		
150	153	+3		
160	163	+3		
170	174	+4		
180	184	+4		
190	194	+4		
200	205	+5		
220	225	+5		
240	246	+6		
260	266	+6		
280	287	+7		
300	308	+8		
350	—			
400	—			

POWER PLANT CHART

AIRCRAFT MODEL	PROPELLER	ENGINE MODEL
F4U-4	BLADE: 6501A-0	R-2800-18W
	HUB: 24E60	CARBURETOR
		BENDIX-STROMBERG PR-58E2

GAGE READING	FUEL PRESS.	OIL PRESS.	OIL TEMP.
DESIRED	17	85	75°C (167°F)
MAXIMUM	18	100	100°C (212°F)
MINIMUM	16	70	
IDLING	9	25	

MAXIMUM PERMISSIBLE DIVING RPM: 3120 (30 SEC.)
MINIMUM RECOMMENDED CRUISE RPM: 1200
OIL GRADE: 1100, SPEC. AN-VV-O-446
FUEL GRADE: GRADE 100/130, SPEC. AN-F-28

COMBAT POWER (3) (COMBAT EMERGENCY)			MILITARY POWER (NON-COMBAT EMERGENCY)			OPERATING CONDITION			NORMAL RATED (MAX. CONTINUOUS)			MAXIMUM CRUISE (NORMAL OPERATION)		
5 MINUTES 245°C (473°F)			5 MINUTES 245°C (473°F)			TIME LIMIT MAX. CYL. HD. TEMP.			UNLIMITED 232°C (450°F) OR 1 HOUR 245°C (473°F)			UNLIMITED 232°C (450°F)		
AUTO RICH 2800			AUTO RICH 2800			MIXTURE R.P.M.			AUTO LEAN 2600			AUTO LEAN 2150 NEUTRAL 2150 LOW BLOWER 2100 HIGH BLOWER		
MANIF. PRESS.	SUPER- CHARGER	FUEL(1) GAL./MIN.	MANIF. PRESS.	SUPER- CHARGER	FUEL(1) GAL./MIN.	STD. TEMP. °C	PRESSURE ALTITUDE	STD. TEMP. °F	MANIF. PRESS.	SUPER- CHARGER	FUEL(2) GPH	MANIF. PRESS.	SUPER- CHARGER	FUEL(2) GPH
						−55	40000	−67						
						−55	38000	−67	30.5 F.T.	H	121	19 F.T.	H	35
			37 F.T.	H	3.3	−55	36000	−67	34 F.T.	H	142	21.5 F.T.	H	41
			41 F.T.	H	3.6	−52.4	34000	−67.3	37.5 F.T.	H	164	23.5 F.T.	H	48
43.5 F.T	H	3.6	45 F.T.	H	4.0	−48.4	32000	−55.1	41 F.T.	H	187	26 F.T.	H	54
47.5 F.T.	H	3.8	49 F.T.	H	4.4	−44.4	30000	−48	44.5 F.T.	H	209	28.5 F.T.	H	60
51.5 F.T.	H	4.1	53 F.T.	H	4.7	−40.5	28000	−40.9	47.5	H	231	31 F.T.	H	67
56 F.T.	H	4.4	54	H	4.8	−36.5	26000	−33.7	47.5	H	231	34 F.T.	H	74
60	H	4.9	54	H	4.8	−32.5	24000	−26.5	47.5	H	231	36.5 F.T.	H	81
60	H	4.9	52 F.T.	L	4.8	−28.6	22000	−19.4	46.5	L	218	37.5	H	83
60	H	4.9	54	L	4.9	−24.6	20000	−12.3	48	L	224	35.5 F.T.	L	84
57 F.T.	L	4.4	54	L	4.9	−20.7	18000	−5.2	48	L	224	37	L	88
60	L	4.7	54	L	4.9	−16.7	16000	2.0	48	L	224	37	L	88
60	L	4.7	54	L	4.9	−12.7	14000	9.1	48	L	224	37	L	88
60	L	4.7	54	L	4.9	−8.8	12000	16.2	48	L	224	37	L	88
60	L	4.7	54	L	4.9	−4.6	10000	23.4	48	L	224	31.5	N	90
60	L	4.7	54	L	4.9	−0.8	8000	30.5	48	L	224	31.5	N	90
60	L	4.7	54	L	4.9	3.1	6000	37.5	43.5	N	188	31.5	N	90
60	L	4.7	51 F.T.	N	4.5	7.1	4000	44.7	43.5	N	188	31.5	N	90
60	L	4.7	54.5	N	4.6	11	2000	51.8	43.5	N	188	31.5	N	90
60	N	4.6	54.5	N	4.6	15	SEA LEVEL	59	43.5	N	188	31.5	N	90

GENERAL NOTES

(1) GAL/MIN: APPROXIMATE U.S. GALLON PER MINUTE PER ENGINE.
(2) GPH: APPROXIMATE U.S. GALLON PER HOUR PER ENGINE.
F.T. MEANS FULL THROTTLE OPERATION. CRITICAL AND BLOWER SHIFT ALTITUDES ARE APPROX. AND ARE BASED ON NO-RAM STANDARD CONDITIONS.
(3) COMBAT POWER, FORMERLY TERMED WAR EMERGENCY POWER.

FOR COMPLETE CRUISING DATA SEE APPENDIX I.
NOTE: TO DETERMINE CONSUMPTION IN BRITISH IMPERIAL UNITS, MULTIPLY BY 10 THEN DIVIDE BY 12. RED FIGURES ARE PRELIMINARY, SUBJECT TO REVISION AFTER FLIGHT CHECK.

TAKE-OFF CONDITIONS: 54.5 IN. HG. MAN. PRESS., 2800 RPM, AUTO-RICH, NEUTRAL BLOWER

CONDITIONS TO AVOID:

SPECIAL NOTES

DATA AS OF: AUGUST 29, 1944

*ENGINE DRIVEN FUEL PUMP ONLY, WITH BOTH ENGINE DRIVEN PUMP AND AUXILIARY FUEL PUMP IN OPERATION, "EMERGENCY" SETTING, FUEL PRESSURE MAY BE SLIGHTLY HIGHER.

RESTRICTED
AN-01-45HB-1

Section IV
Paragraphs 1-2

Section IV
EMERGENCY OPERATING INSTRUCTIONS

1. EMERGENCY EGRESS.

a. The entire cabin sliding section may be released in case of emergency. The two release handles, one on either side of the cabin structure, are plainly marked "CABIN EMERGENCY RELEASE." These two handles are safety-pinned to prevent inadvertent release of the cabin. The safety pins (painted RED) are attached to wire loops adjacent to the release handles and must be pulled free (aft) before the release handles can be moved. The release handles disengage the front rollers from the cabin. As the cabin is pushed upward, the rear rollers are disengaged and the cabin is freed from the airplane to be carried away by the slipstream. (See figure 24.)

b. To release the cabin in an emergency in flight:

(1) Pull the safety pin loops.

(2) Pull both cabin release handles inboard and push forward.

(3) Push upward on the release handles to break the cabin free.

2. EMERGENCY LANDING GEAR OPERATION.

a. In case of failure of the hydraulic system or

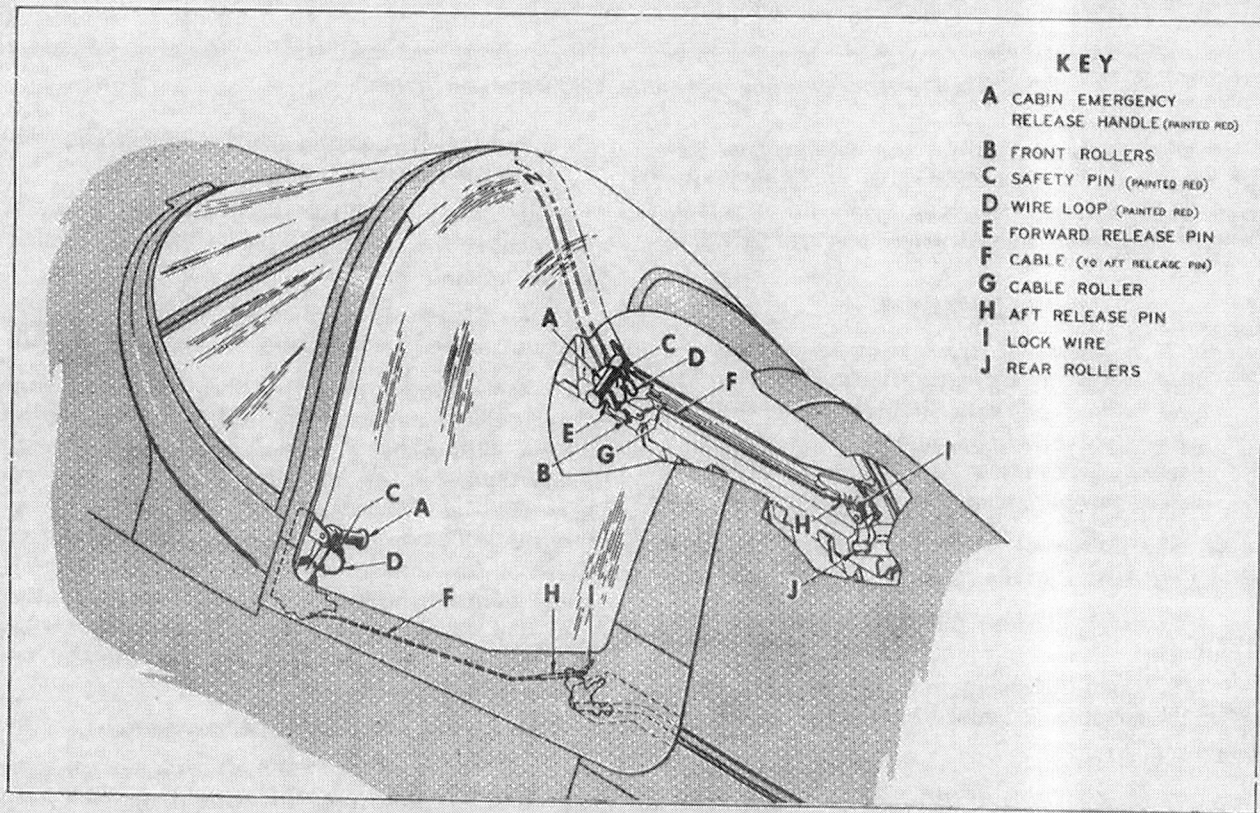

KEY

- **A** CABIN EMERGENCY RELEASE HANDLE (PAINTED RED)
- **B** FRONT ROLLERS
- **C** SAFETY PIN (PAINTED RED)
- **D** WIRE LOOP (PAINTED RED)
- **E** FORWARD RELEASE PIN
- **F** CABLE (TO AFT RELEASE PIN)
- **G** CABLE ROLLER
- **H** AFT RELEASE PIN
- **I** LOCK WIRE
- **J** REAR ROLLERS

Figure 24 — Cabin Emergency Release

RESTRICTED

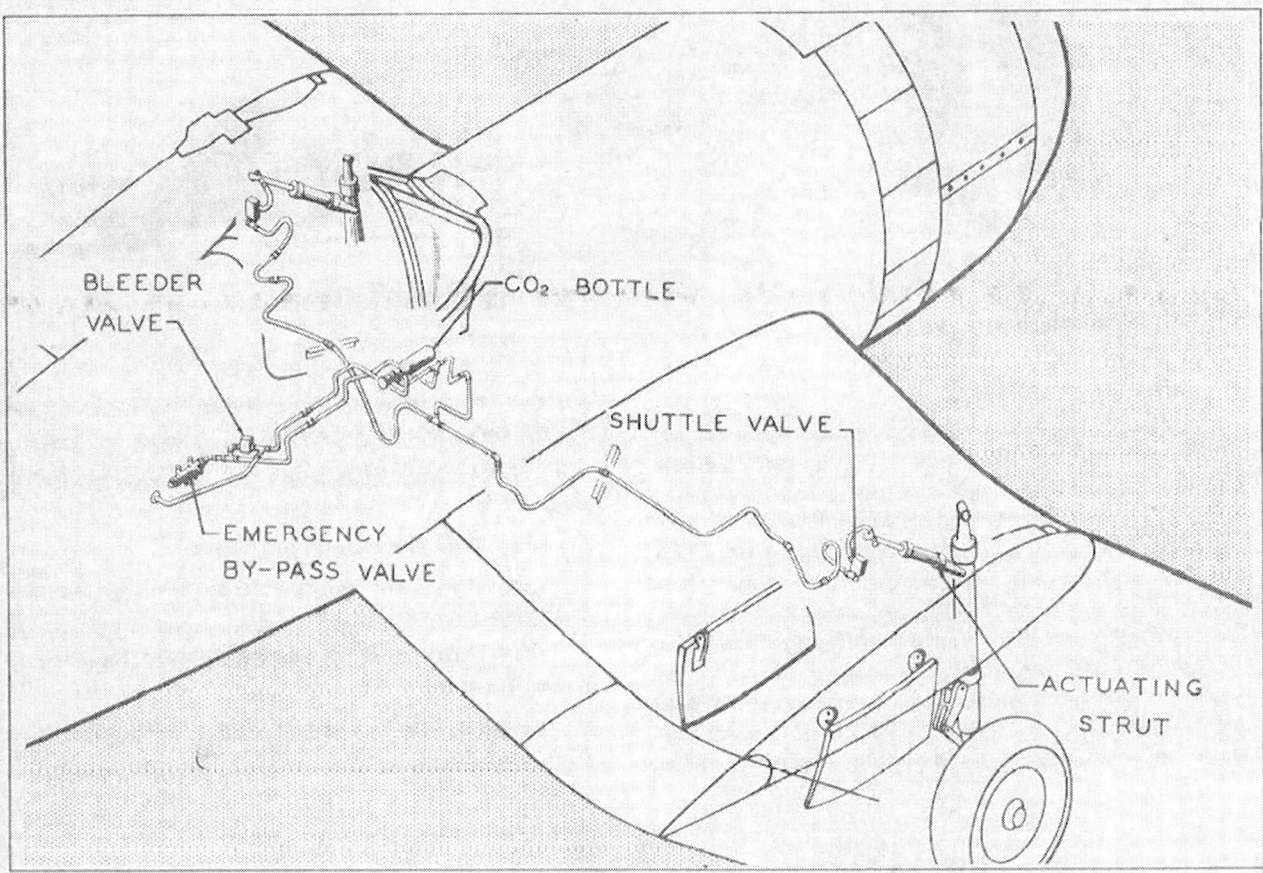

Figure 25 — Emergency Landing Gear Extension System

even complete power failure, the landing gear can be extended by means of the CO_2 emergency system. This acts upon the main gear while a spring system acts upon the tail wheel (see figure 25).

WARNING

If it is suspected that a suitable landing platform, such as a carrier deck or landing field, is not available, it must be remembered that it will be impossible to again retract the landing gear after the CO_2 emergency extension system has been used.

b. The following procedure is used for emergency extension of the landing gear.

(1) Close throttle and reduce speed to about 110 knots.

(2) Open the emergency landing gear release valve. This valve is located on the center control panel.

Note

The CO_2 system will extend the landing gear regardless of the position of the landing gear control handle.

(3) Further reduce speed to about 90 knots (keep above the stalling speed) while the landing gear is extending.

(4) Check the indicators that the landing gear and tail wheel are fully locked "DOWN."

c. The emergency extension of the landing gear is started at a comparatively high speed so that the airflow will assist in opening the landing gear doors. Opening the emergency landing gear release valve admits CO_2 to a shuttle valve and a by-pass valve. The by-pass valve by-passes the hydraulic oil at the bottom of the landing gear and tail wheel struts directly back to the hydraulic reservoir. The shuttle valve in turn admits CO_2 pressure to the top of the landing gear struts, thereby extending the gear.

3. EMERGENCY WING FLAP OPERATION.

a. In the event that the hydraulic system fails, provision for lowering the wing flaps has been made. Pull the emergency wing flap selector, located on the left hand shelf, aft from its normal "HYDRAULIC SYSTEM" position to "EMERGENCY FLAPS." Pressure from the hand pump will then by-pass the main pressure line and operate the wing flaps directly. See figure 3.

4. LIFE RAFT.

a. The one-man parachute-type life raft is used by pilots operating this airplane.

5. ENGINE FAILURE DURING FLIGHT.

a. Engine failure may be indicated by either of the following symptoms:

(1) Freezing of the engine.

(2) Drop in altitude and loss of speed.

Note

If the engine fails but does not freeze, no absence of engine noise is apparent since the windmilling propeller simulates normal engine operation. Also, in this condition, manifold pressure can be increased and decreased normally, and the propeller blade angle can be changed within certain limits. While the propeller is windmilling, the hydraulic system can be operated normally. However, if the engine should freeze or rough operation should necessitate stopping the engine by placing the propeller governor control in high pitch (full "DECREASE") position, the hydraulically controlled units must be operated by the hand pump.

b. If altitude permits, attempt to find the cause of engine failure as follows:

(1) The selected tank may be empty. Switch to another tank.

(2) If it is apparent that the fault does not lie in fuel system operation, and altitude still permits, check the following:

(a) Move the mixture control to "AUTO RICH."

(b) Test the magnetos individually.

c. If, after completing the above operations, the engine does not start, prepare for an emergency landing. Refer to paragraph 6., below.

Note

The gliding ratio of this airplane in the clean condition at 140 knots indicated air speed (best gliding speed) is 13:1.

6. FORCED LANDINGS.

a. GENERAL.

(1) In the event of a forced landing over land, the pilot should consider a number of variables in order to determine his best landing attitude. These include altitude, type of terrain, and the characteristics of the airplane.

(2) Landings in terrain such as golf courses, ploughed fields, swamps, mud or sand should be made with landing gear up. Most nose-overs occur as a result of landing in such territory with the landing gear down, and nearly all serious injuries and fatalities result from nosing over.

(3) Landings in rough, rocky or tree stump terrain should be made with wheels down so that the undercarriage and not the fuselage will make the initial contact.

(4) Pilots should remember that ground which appears smooth and level from the air frequently turns out to be rough, crossed with ditches, soft, or full of obstructions when the actual landing is made.

(5) All forced landings should be made well above the stalling speed. There will be little or no control of the airplane if an attempt is made to land at or slightly above stalling speed.

WARNING

In the event of a forced landing with drop tank or bombs attached, release in accordance with instructions contained in Section II, paragraph 4. *b.* (3).

b. BELLY LANDINGS.

(1) PREPARATION FOR BELLY LANDINGS.

(a) Release drop tank or bombs.

(b) Landing Gear— "UP."

(c) Landing flaps—"DOWN."

(d) Shoulder harness and safety belt—locked tight.

(e) Jettison the cockpit cabin sliding section.

(f) Fuel tank pressure release—"DUMP" (forward).

(2) PRIOR TO CONTACT WITH THE GROUND:

(a) Drop pilot's seat several inches.

(b) Switches (battery, generator, ignition) "OFF."

(c) Fuel selector—"OFF."

**Section IV
Paragraphs 6-8**

c. WATER LANDINGS (DITCHING).

(1) The same procedure as that outlined for belly landings is applicable to ditching.

Note

THIS AIRPLANE HAS EXCELLENT WATER LANDING CHARACTERISTICS DUE TO THE INVERTED GULL WING WHICH CAUSES IT TO PLANE ON CONTACT WITH THE WATER. BECAUSE OF THE PLANING FEATURE, A FULL-STALL LANDING IS NOT NECESSARY.

7. ELECTRICAL FIRE—In the event of a fire in the electrical system, the following procedure should be applied:

a. Turn off the emergency generator and battery switches.

b. Turn off all other electrical switches.

c. If the fire is extinguished, turn the circuits on one at a time, starting with the emergency generator and battery switches and watching the circuit which caused the fire.

8. GENERATOR SYSTEM FAILURE.

a. HIGH VOLTAGE (over 30.0).—The following steps should be taken to prevent burning out the battery and other equipment:

(1) Turn the emergency generator switch to the "OFF" position.

(2) Operate only absolutely essential loads from the battery.

(3) The battery can be periodically recharged by closing the emergency generator switch for not more than five minutes. During these charging intervals, the electrical equipment likely to be damaged by excessive voltage should be turned off.

b. LOW VOLTAGE (below 26.0) may permit the reverse current cut-out to open and allow the electrical equipment to drain the battery power. Loss of generator power and operation on battery power may be detected by failure of the radio equipment to operate satisfactorily.

(1) To check whether the reverse current cut-out is closed, open the battery switch and observe whether electrical loads (radio, lights, instruments, etc.) lose electrical power. If they do lose power, then the reverse current cut-out is open, and all but absolutely essential electrical loads should be turned off in order to conserve the battery power.

c. VERY LOW OR ZERO VOLTAGE.—Promptly turn off all but absolutely essential loads to conserve the battery.

Hey! You're out of the 98 cu. in. class!

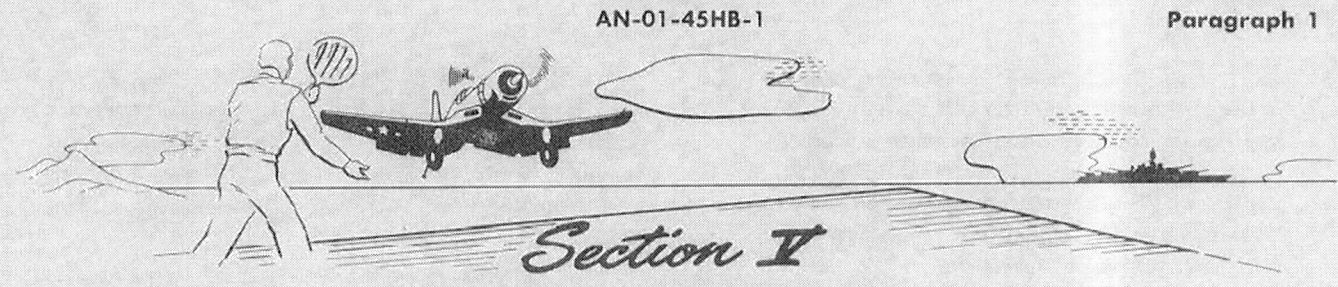

Section V

OPERATIONAL EQUIPMENT

1. OPERATION OF OXYGEN EQUIPMENT.

a. GENERAL.—A diluter-demand type oxygen system is provided (see figure 26). The oxygen bottle is located underneath the floor of the cockpit, the main valve on the bottle protruding through the cockpit floor just inboard of the right hand shelf. The oxygen bottle refill valve is accessible through the cockpit access door in the bottom of the fuselage. The oxygen supply may be replenished by replacing a depleted cylinder with a full cylinder or by refilling the cylinder installed through the filler valve. (This arrangement precludes the necessity of removing the bottle for refilling.) The diluter-demand regulator is located above the forward end of the left hand shelf. The diluter-demand regulator is similar in operation to the demand regulator except that an air admission valve, which allows air from the outside to enter the breathing system, is incorporated. The amount of air admitted is dependent upon the altitude up to approximately 30,000 feet, beyond which 100 per cent oxygen is automatically delivered.

b. WHEN TO USE OXYGEN.

(1) During normal operations the diluter lever should be turned to the "ON" position, thus obtaining the maximum economy and endurance from the oxygen supply aboard.

Note

If symptoms suggestive of oxygen defici-

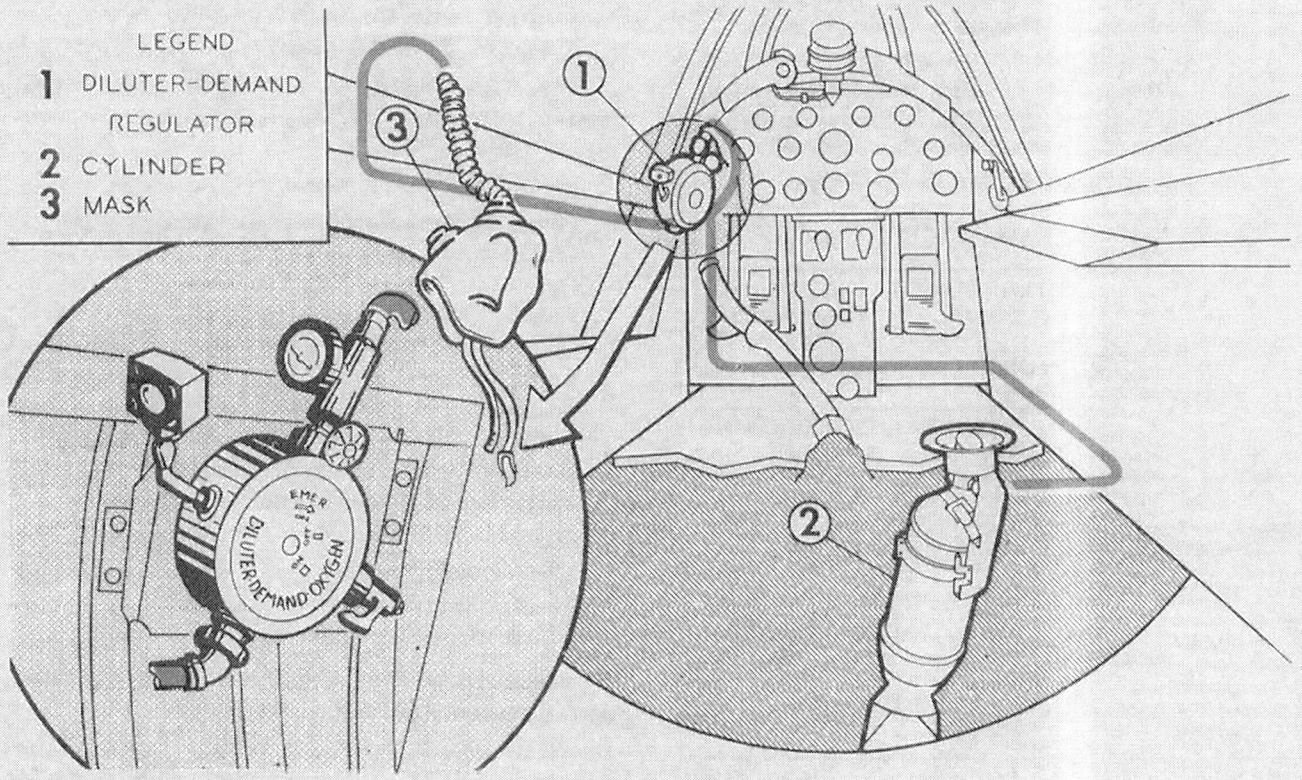

Figure 26 — Oxygen System

ency (anoxia) such as drowsiness, dizziness, dimming of vision, awkward performance of routine tasks, or nausea should occur, descend immediately to 10,000 feet, using the emergency oxygen supply, and inspect the oxygen system in accordance with paragraph 1.c., below.

(2) The emergency valve (small red knob on regulator) shall be used only if the diluter-demand regulator becomes inoperative or if anoxia is suspected. The regulator is equipped with an oxygen flow indicator which varies with the pilot's breathing, thereby indicating proper operation of the system. If the blinking action ceases, immediately turn on the emergency valve; open the emergency valve slowly and obtain the minimum flow required.

(3) Use oxygen on all flights above 10,000 feet.

(4) On all flights of more than four hours between 8,000 and 10,000 feet, oxygen shall be used a minimum of 15 minutes out of every hour.

(5) Use oxygen on night flights above 5,000 feet.

Note

When carrying a drop tank with main tank full, it is possible to operate for a period longer than the oxygen supply will last. Study the Oxygen Consumption Table below and the Flight Operation Instruction Charts in Appendix I carefully and plan oxygen flights accordingly.

OXYGEN CONSUMPTION TABLE

ALTITUDE IN FEET	DILUTER-DEMAND REGULATOR WITH DILUTER "OFF" ENDURANCE HOURS	DILUTER-DEMAND REGULATOR WITH DILUTER "ON" ENDURANCE HOURS
5,000	1.8	7.0
10,000	2.1	8.3
15,000	2.6	10.0
20,000	3.3	8.8
25,000	4.1	6.0
30,000	5.0	5.0
35,000	6.5	6.5
40,000	8.3	8.3

c. PRE-FLIGHT CHECK LIST.—The following items shall be checked at regular intervals when the airplane is on the ground, and whenever possible before flights in which oxygen is likely to be used, to assure proper functioning of the oxygen system:

(1) Check regulator emergency valve to determine that it is closed.

(2) Open cylinder valve. Allow at least ten seconds for pressure in line to equalize. Pressure gage should read 1800 ±50 pounds per square inch if the cylinder is fully charged.

(3) Close cylinder valve. If pressure drops more than 100 pounds in five minutes, there is excessive leakage. The system should be repaired prior to use.

(4) Check mask fit by squeezing off the corrugated breathing tube and inhaling lightly. The mask will adhere tightly to the face if there is no leakage. DO NOT USE A MASK THAT LEAKS. Never check mask fit, as outlined, with EMERGENCY FLOW "ON."

(5) Couple mask securely to breathing tube by means of quick disconnect coupling.

Note

Mating parts of coupling must not be "cocked", but fully engaged.

(6) Open cylinder valve. Breathe several times to determine whether the regulator is functioning properly.

Note

Since the amount of added oxygen is very small at sea level, the oxygen flow meter may not operate while the plane is on the ground. In this case, turn the air valve to "OFF" or "100% OXYGEN" and test again. If oxygen flow indicator operation is now satisfactory, reset the air valve to "ON" or "NORMAL OXYGEN," in which setting adequate oxygen flow and "blinker" operation will be assured at oxygen-use altitudes.

(7) Check the emergency valve by turning the handle toward the "ON" position until oxygen flows into the mask. Close the emergency valve.

2. OPERATION OF RADIO, COMMUNICATION AND NAVIGATION EQUIPMENT.

a. GENERAL.—This airplane is furnished with AN/ARC-1 communication, AN/ARR-2A navigation and AN/APX-1 identification radio and radar equipment.

Figure 27 — Radio and Communications Controls

(1) AN/ARC-1 COMMUNICATION—RECEIVING.

(a) EQUIPMENT.

Receiver	Frequency
RT-18/ARC-1	100 to 156 kilocycles
R-23/ARC-5	190 to 550 kilocycles
R-26/ARC-5	3.0 to 6.0 megacycles

(b) OPERATION.

1. Plug headphone and microphone plugs into the pilot's jack box, located to the right and aft of the pilot.

2. Check battery switch; must be in "ON" position.

3. On MASTER panel: throw master radio switch to "ON" position. (It will take about one minute to warm up.)

4. On MIXER panel: under "RECEIVER OUTPUT" push forward either the HF or VHF switch, depending on the particular pre-tuned frequency desired.

a. If VHF is chosen: on VHF panel:

(1) Turn CHAN SEL switch to the desired main channel. (Allow at least 20 seconds for warm-up.)

(2) Turn GUARD-BOTH-MAIN T/R switch to "BOTH." (This is the position for normal operation.)

(3) Turn GUARD-BOTH-MAIN T/R switch to "GUARD" or "MAIN T/R," depending on which of the desired signals are recognized.

Note

Reception may be prevented by interfering signals or noise passed by the other channel. Suppress this by turning to "MAIN T/R" or "GUARD" as required.

b. If the HF receiver is chosen: on MIXER panel, adjust sensitivity control to an agreeable noise level.

5. On MASTER panel: turn COMM VOLUME knob to the desired intensity.

6. To operate LF ferry receiver:

a. On RECVR panel turn tuning knob to obtain the desired dial frequency reading.

b. Turn SENS knob clockwise to the desired volume intensity.

7. To turn off equipment:

a. VHF: On VHF panel turn CHAN SEL switch to "OFF" position.

b. HF: On MIXER panel, throw HF switch to "OUT" position.

c. LF ferry receiver: On RECVR panel turn SENS knob fully counterclockwise.

Note

LF ferry receiver is on whenever the master radio switch is on.

(2) AN/ARC-1 COMMUNICATION — TRANSMITTING.

(a) EQUIPMENT.

Transmitter	Frequency
RT-18/ARC-1	100 to 156 megacycles
T-19/ARC-5	3 to 4 megacycles

(b) OPERATION.

1. Plug headphone and microphone plugs into pilot's jack box.

2. Check battery switch; must be in "ON" position.

3. On MASTER panel: throw master radio switch to "ON" position. (It will take about one minute to warm up.)

4. On MIXER panel: turn MIC SEL indicator to either HF or VHF, depending on the particular frequency desired.

a. If VHF is selected: on the VHF panel:

(1) Turn CHAN SEL switch to the desired main channel. (Allow at least 20 seconds for warm up.)

(2) Turn GUARD-BOTH-MAIN T/R switch to "GUARD" or "MAIN T/R," depending on the frequency desired.

(3) Proceed as outlined in par. 5 below.

b. If HF is chosen, nothing further is necessary except to proceed as outlined below.

5. Microphone selection and operation.

a. If hand microphone is used, depress button on microphone to talk.

b. If mask microphone is needed, push the throttle switch button.

c. Hold microphone as close to the lips as possible and speak clearly and distinctly. It is not necessary to shout.

6. To turn off equipment:

a. Depending on the type of microphone used, release either the hand microphone button or the throttle switch button for the mask microphone.

(3) AN/ARR-2A NAVIGATION.

(a) EQUIPMENT.

Receiver	Frequency
R-4A/ARR-2A	234 to 258 megacycles

(b) OPERATION.

1. On NAVIG panel: See that VOICE-NAV indicator is in NAV position.

2. Set CHAN SEL indicator to the particular channel desired of the six channels available.

3. Turn SENS knob clockwise to obtain a comfortable volume intensity.

4. Vary pitch (or beat note) by turning PITCH knob to a desirable frequency.

5. To turn off equipment:

a. Turn SENS knob fully counterclockwise.

(c) NAVIGATION ANTENNA. — The control for extending and retracting this antenna is located on the right side of the cockpit, just below the cabin track. The antenna is extended by unlatching, pulling the handle aft and latching. The antenna should be extended only when actually being used in flight, since it causes a certain definite, though small, loss in maximum speed (1 mph).

(4) AN/APX-1 IDENTIFICATION.

(a) OPERATION.

Note

Before take-off check that a complete destructor circuit test has been made.

1. On IFF panel: set CODE indicator to the desired position of the six positions available. (Set to position No. 1 if no other has been previously specified.)

2. Operate G BAND switch as required.

3. To turn off equipment:

a. Be certain that G BAND switch is in "OUT" position and CODE indicator is in "OFF" position.

Note

Additional information concerning operation of identification equipment should be obtained from the communications officer in charge.

3. OPERATION OF ELECTRICAL EQUIPMENT.

a. GENERAL.—The electric power for the airplane is supplied by a 28-volt system in which the framework of the airplane acts as the ground return. The power may be provided by one of three sources, the 75 ampere generator, the battery, or the external power receptacle. To a large extent, the electrical system is controlled from the elec-

RESTRICTED
AN-01-45HB-1

Section V
Paragraph 3

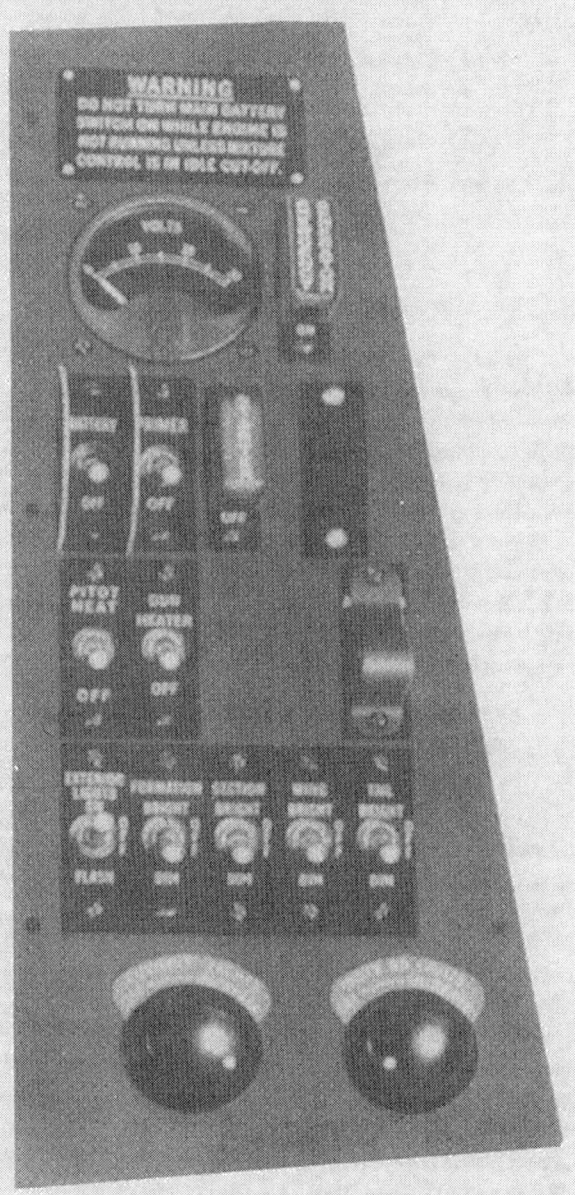

Figure 28 — Electrical Controls

trical control box on the right hand side of the cockpit. See figure 28.

(1) BATTERY SWITCH. — The battery switch, located on the electrical control box, must be "ON" in order that current may be supplied from the battery. When the battery switch is "OFF," the battery cannot supply power to any external load except the IFF destructor (inertia) switch. When the airplane is on the ground with the engine off, the battery switch should be "OFF."

This will prevent any inadvertent drain on the battery.

(a) For all ground running and flight operation, the battery switch should be turned to "ON." The power for starting the engine may be taken either from the battery or from an external power source.

Note

When power is taken through the external power receptacle, the battery switch should be in the "OFF" position.

WARNING

Do not turn the battery switch "ON" while the engine is not running unless the mixture control is in "IDLE CUT-OFF." If the fuel pump switch is turned to "BOOST" or "EMERGENCY" while the mixture control is in any position but "IDLE CUT-OFF" and the battery switch is "ON," the lower cylinders may become flooded, resulting in damage to the engine when it is subsequently started.

(2) EMERGENCY GENERATOR SWITCH.—The emergency generator switch, located on the electrical control box, is normally closed. Power for the electric system is thus obtained from the generator, regulated by the voltage regulator, and measured by a voltmeter. The voltmeter, located on the top forward section of the electrical control box, should read between 27.5 and 28.5 volts.

(a) A faulty generator, voltage regulator, or battery will be indicated by an erratic or improper reading on the voltmeter. Open the emergency generator switch; this will isolate the generator from the remainder of the electrical system. If the fault is not with the battery, it will automatically take over the load as long as the battery switch is on. All electrical equipment unessential to flight should be turned off. When two batteries are installed, radio and IFF operation will last about one-half hour running on the batteries alone.

(b) A generator cutout access hole is located just above the floor on the electrical control box. If the voltmeter shows no reading, insert a finger through the hole and push the armature of the cutout to close the circuit and bring the generator back into the system.

(3) MOMENTARY CONTACT SWITCHES.—The starter, primer, and oil dilution switches are momentary contact switches, viz., the switch

RESTRICTED

49

must be held in the "ON" position until the operation is completed. These switches are located on the electrical control box.

(4) EXTERIOR LIGHT MASTER SWITCH.—The three-position exterior light master switch is located on the electrical control box. The positions are "ON," "OFF" and "FLASH." The "FLASH" position is momentary and can be used as a keying switch. The switch must be turned "ON" or to "FLASH" in order to energize the individual exterior light switches, viz., the formation light switch, the section light switch, the wing light switch, and the tail light switch. These switches also have three positions, "BRIGHT," "OFF," and "DIM."

Note

When operating in combat areas, the exterior lights can be pre-selected prior to take-off and turned on by turning the master switch to "ON" after the take-off has been completed.

(5) COCKPIT LIGHTS.—A single rheostat, located on the electrical control box, controls all of the cockpit lights. Each light has a built-in switch so that individual selection can be made.

(6) INSTRUMENT BOARD LIGHTS.—The instrument board lights are controlled by a rheostat which is located next to the cockpit light rheostat on the electrical control box.

Note

All lights in the cockpit are "RED" in order to disturb vision in night flying as little as possible. (See Section II, paragraph 3.b.)

(7) CIRCUIT BREAKERS.—All of the circuits in the airplane are protected by circuit breakers located on the vertical side panel of the electrical control box. The circuit breakers are designed to maintain a closed circuit up to the rated current for the circuit. If an electrical overload of sufficient magnitude and duration occurs in a circuit, the circuit breaker button will pop out, thus breaking the circuit. Push the button back in; if the circuit has been seriously disturbed, as by a "short," the button will pop out again, breaking the circuit.

(8) SWITCHES NOT LOCATED ON ELECTRICAL CONTROL BOX.—In addition to the switches located on the armament switch boxes above the instrument panel, there are five electrical switches located in parts of the cockpit other than the electrical control box. They are located as follows:

(a) Center control panel.
1. Oil cooler flap control switch.
2. Intercooler flap switch.

(b) Left hand shelf.
1. Cowl flap control switch.
2. Booster fuel pump switch.
3. Transfer pump switch.

(9) APPROACH LIGHT.—The approach light is located in the leading edge of the left outer panel. For carrier-based operations a fixed guard is placed over the approach light switch (located on the electrical control box) holding the switch in the "OFF" position, the approach light being turned on automatically by the extending arresting hook. The switch guard may be removed for land-based operations, permitting operation of the approach light without extension of the arresting hook.

4. OPERATION OF ARMAMENT.

a. GUNS.

(1) GENERAL.—The airplane is armed with six .50 caliber Browning machine guns in the outer panels. There are six ammunition boxes, two per gun, in each outer panel, supplying 400 rounds of ammunition to each inboard and intermediate gun and 375 rounds to each outboard gun. The guns are charged hydraulically and fired electrically. The controls include the gun charging knobs on the center control panel, the master armament switch, the individual switches for each pair of guns (inboard, intermediate, and outboard), all located on the armament switch boxes above the instrument panel, and the trigger switch on the control stick. To fire the guns, turn the master armament switch to either "BOMBS and GUNS" or "ROCKETS and GUNS" and the individual gun switches to "ON"; press the trigger switch. The guns will fire as long as the trigger switch is closed.

Note

If the trigger switch sticks and the trouble cannot be remedied easily, uncontrolled automatic firing will take place; turn off the master armament switch and use it as a trigger. In order to stop a "runaway" gun, depress the proper charging knob and turn it to "SAFE."

(2) GUN SIGHT.—The Mark 8 illuminated gun sight is located on the cowl deck above the

instrument panel. The gun sight switch and the rheostat for the gun sight light are on the left hand armament switch box. The gun sight switch has three positions, "ON," "OFF," and "ALT." Turn to the "ALT." position (alternate filament) whenever the light goes out. Spare bulbs for the sight are carried above the instrument panel on the left hand side.

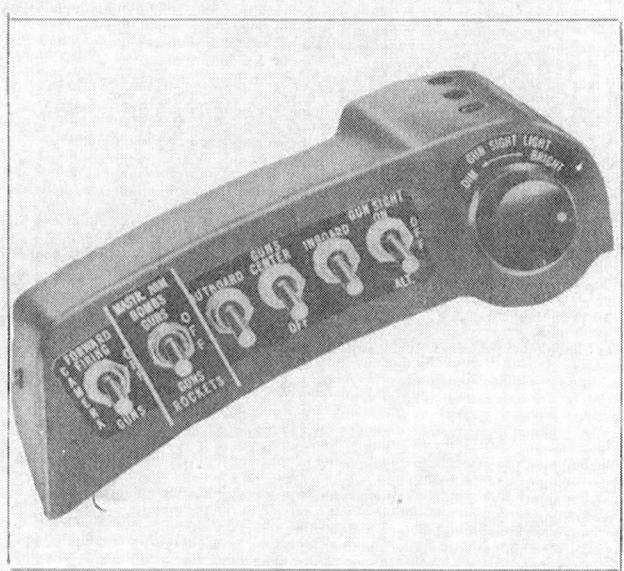

Figure 29 — Gun Switch Box

(3) GUN CAMERA.—The gun camera is installed in the right hand outer panel. The control switch, located on the left hand armament switch box, has the following positions. "FORWARD FIRING," "GUNS," and "OFF." The camera circuit is in series with the gun trigger switch on the control stick and will take motion pictures of either gun or rocket targets without firing the guns or rockets.

(a) Control settings for pictures of gun targets are:

1. Master armament switch — "BOMBS" and "GUNS" or "ROCKETS and GUNS."
2. Gun camera switch — "GUNS."
3. Press the trigger switch.

(b) Control settings for pictures of rocket targets are:

1. Master armament switch — "ROCKETS and GUNS."
2. Gun camera switch — "FORWARD FIRING."
3. Press the bomb release button on the control stick.

Note

The above settings do not include provision for gun firing or rocket launching. For these, either the individual gun switches must be turned on or the rocket safety plug must be inserted respectively.

(4) GUN HEATING.—The gun heating system consists of electrically heated pads attached to each gun. The gun heater switch is located on the electrical and radio control box.

b. BOMBS.

(1) GENERAL.—Provision is made for carrying bombs on the twin pylons. A set of switches located on the right hand armament switch box controls bomb arming and selection of the bomb to be released. The thumb switch for releasing the bombs is located on the control stick.

(2) BOMB ARMING.—The bomb arming switch has three positions, "TAIL," "SAFE," and "NOSE and TAIL." The pilot will select the type of arming he desires. This selection will depend on the use of the bomb. If it is to be armor piercing, fuse the tail of the bomb; if it is to be anti-personnel, fuse both nose and tail for explosion on contact. If the bomb is not armed, it will fall safe and will not explode.

(3) BOMB RELEASE.—Bombs may be released either manually or electrically in the same manner as droppable fuel tanks. Refer to Section II, paragraph 4.b.(3) for complete release instructions.

c. ROCKETS.

(1) GENERAL.—Provision is made for carrying four rockets on each outer wing panel. A set of switches on the right hand armament switch box and on the instrument panel controls arming

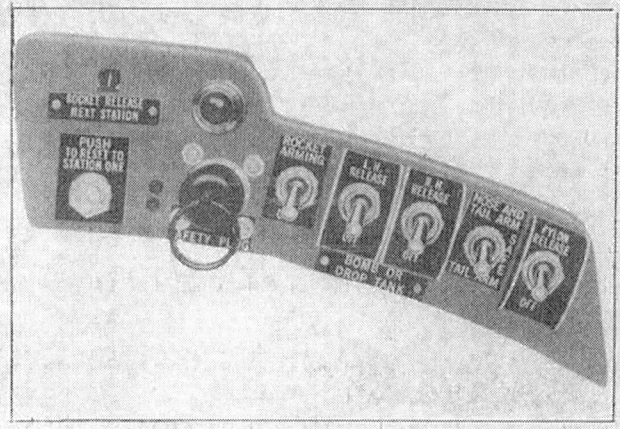

Figure 30 — Rocket and Bomb Switch Box

and firing of the rockets. The rockets are fired by the thumb switch on the top of the control stick.

Note

The camera switch, located on the left hand armament switch box, must be turned to "FORWARD FIRING" in order to operate the camera with the rockets.

(2) ROCKET ARMING. — The arming switch has two positions, "OFF" and "ARM." If the switch is turned to "OFF," the rockets will penetrate before exploding; if turned to "ARM," the rockets will explode on contact. All rockets are armed when the switch is turned to "ARM."

(3) ROCKET FIRING.—The procedure for firing the rockets is as follows:

(a) See that the rocket safety plug is inserted.

(b) Turn the master armament switch to "ROCKETS and GUNS."

Note

When steps (a) and (b) are completed, a red indicator light, located on the right hand armament switch box, will go on.

(c) Arm the rockets, if desired.

(d) Press the bomb release switch (thumb button on the control stick).

(4) ROCKET SAFETY PLUG. — An electrical safety plug is located on the right hand armament switch box. When the plug is removed, the electrical circuits are opened, making it impossible to fire the rockets.

WARNING

The rocket safety plug shall be inserted only in flight. For all take-offs and landings with rockets installed, the safety plug shall be removed.

(5) STATION INDICATING DIAL. — A dial is installed on the right hand armament switch box that indicates the next pair of rockets to be fired. A push button is provided under the dial to reset the dial for each new mission with rockets.

Note

The rockets are fired in pairs, the outboard pairs being fired first. The thumb switch on the control stick must be depressed for each firing.

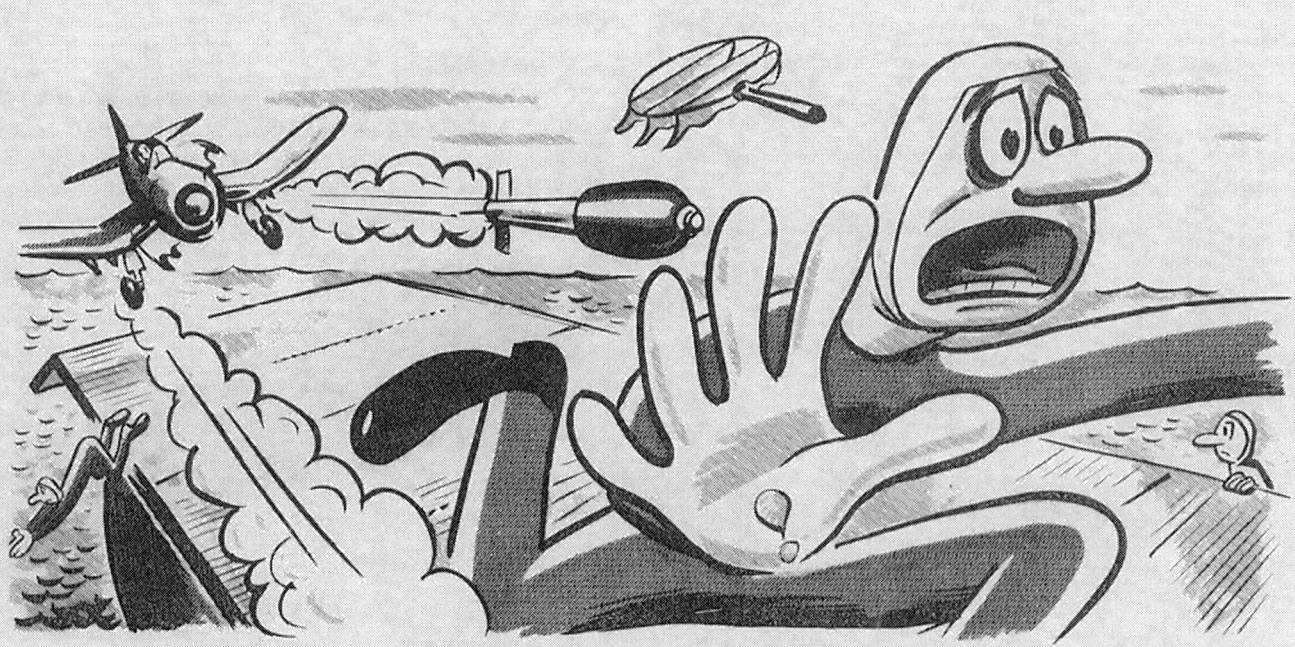

Watch that safety plug!

Appendix

OPERATING CHARTS, TABLES, CURVES AND DIAGRAMS

INDEX

	Page
Figure 31—Protection Against Gunfire	53
Take-off, Climb and Landing Chart	54
Flight Operation Instruction Chart—1	55
Flight Operation Instruction Chart—2	56
Flight Operation Instruction Chart—3	57
Angle of Attack vs. Dive Angle Curves	58
Angle of Attack vs. Indicated Air Speed Curves	59
Engine Calibration Curves	61

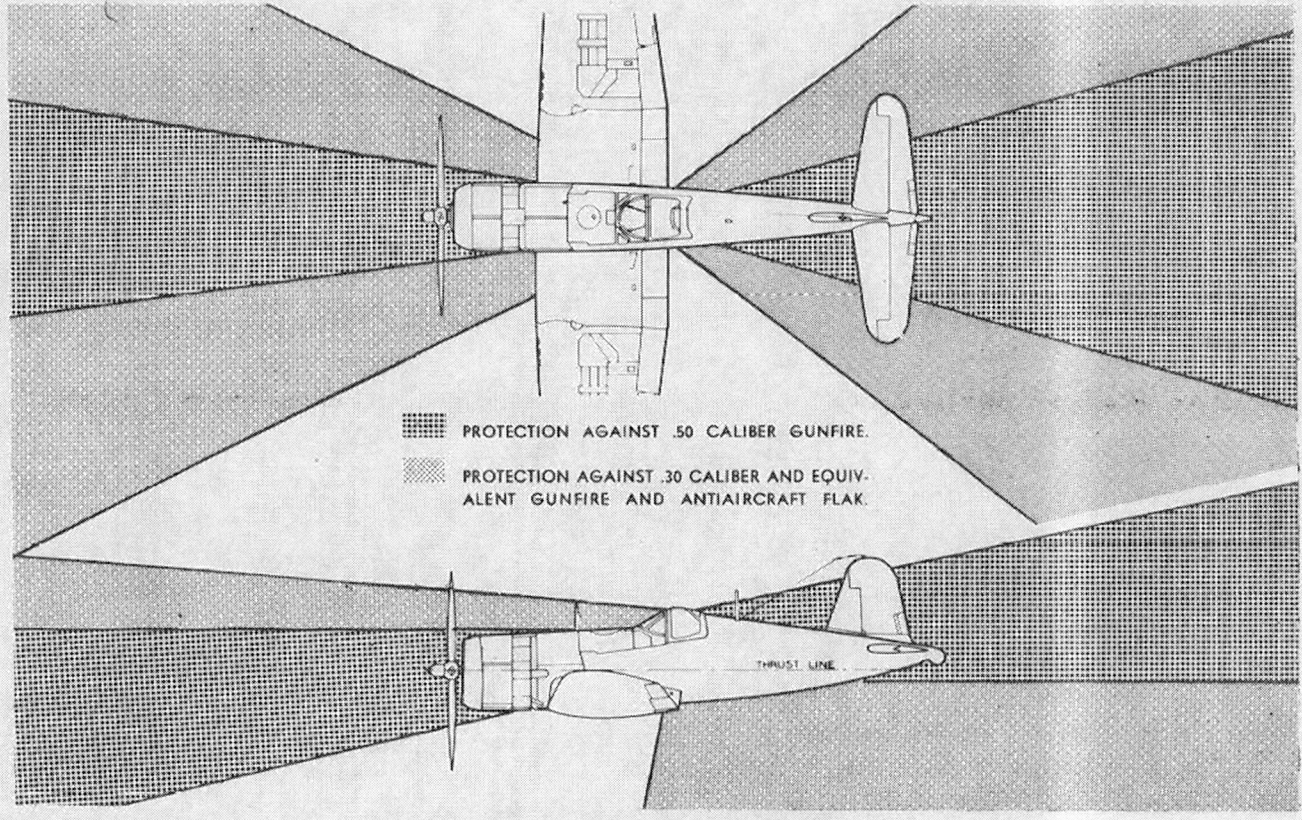

Figure 31 — *Protection Against Gunfire*

Appendix 1 of this publication shall not be carried in aircraft on combat missions or when there is a reasonable chance of its falling into the hands of the enemy.

Appendix I

TAKE-OFF, CLIMB & LANDING CHART

AIRCRAFT MODEL: F4U-4 **ENGINE MODEL:** R-2800-18W

TAKE-OFF DISTANCE FEET

HARD SURFACE RUNWAY

GROSS WEIGHT LB.	HEAD WIND MPH	HEAD WIND KTS	AT SEA LEVEL GROUND RUN	AT SEA LEVEL TO CLEAR 50' OBJ.	AT 3000 FEET GROUND RUN	AT 3000 FEET TO CLEAR 50' OBJ.	AT 6000 FEET GROUND RUN	AT 6000 FEET TO CLEAR 50' OBJ.
12,500	0	0	700	1310	830	1540	1030	1980
	17	15	470	940	570	1120	720	1470
	35	30	290	630	350	750	450	1010
	52	45	130	340	180	430	230	610
13,500	0	0	860	1650	1010	1950	1280	2580
	17	15	590	1200	710	1440	900	1940
	35	30	360	800	440	970	580	1350
	52	45	190	460	240	570	330	840
14,500	0	0	1030	2030	1220	2420	1530	3160
	17	15	720	1500	860	1810	1180	2460
	35	30	440	1010	540	1240	730	1760
	52	45	240	600	310	720	430	1100

SOD TURF RUNWAY

GROSS WEIGHT LB.	HEAD WIND MPH	HEAD WIND KTS	AT SEA LEVEL GROUND RUN	AT SEA LEVEL TO CLEAR 50' OBJ.	AT 3000 FEET GROUND RUN	AT 3000 FEET TO CLEAR 50' OBJ.	AT 6000 FEET GROUND RUN	AT 6000 FEET TO CLEAR 50' OBJ.
12,500	0	0	730	1340	860	1570	1080	2030
	17	15	490	960	590	1140	750	1510
	35	30	300	640	360	770	470	1030
	52	45	140	350	190	440	260	620
13,500	0	0	900	1690	1070	2000	1350	2660
	17	15	610	1230	740	1470	960	2000
	35	30	370	820	460	990	610	1390
	52	45	200	470	260	580	350	860
14,500	0	0	1080	2080	1290	2490	1620	3330
	17	15	750	1640	910	1860	1170	2530
	35	30	470	1030	580	1280	780	1780
	52	45	250	610	320	780	450	1130

SOFT SURFACE RUNWAY

GROSS WEIGHT LB.	HEAD WIND MPH	HEAD WIND KTS	AT SEA LEVEL GROUND RUN	AT SEA LEVEL TO CLEAR 50' OBJ.	AT 3000 FEET GROUND RUN	AT 3000 FEET TO CLEAR 50' OBJ.	AT 6000 FEET GROUND RUN	AT 6000 FEET TO CLEAR 50' OBJ.
12,500	0	0	810	1420	970	1680	1230	2180
	17	15	550	1020	660	1220	860	1610
	35	30	340	670	400	810	540	1090
	52	45	160	360	210	460	300	660
13,500	0	0	1010	1800	1230	2160	1580	2890
	17	15	690	1310	840	1590	1120	2160
	35	30	420	860	530	1070	710	1490
	52	45	220	490	290	620	410	920
14,500	0	0	1240	2240	1530	2700	1940	3640
	17	15	860	1740	1050	2010	1390	2760
	35	30	530	1100	670	1310	910	1930
	52	45	290	640	380	840	540	1210

NOTE: INCREASE CHART DISTANCES AS FOLLOWS: 75°F + 10%, 100°F + 20%, 125°F + 30%, 150°F + 40%.

OPTIMUM TAKE-OFF WITH 2800 R.P.M., 54.5 IN. HG & 50 DEG. FLAP IS 80% OF CHART VALUES

POWER PLANT SETTINGS: (DETAILS ON POWER PLANT CHART, SECTION III)
NOTE: INCREASE ELAPSED CLIMBING TIME 6% FOR EACH 10°C (20°F), ABOVE 15°C (59°F.) FREE AIR TEMPERATURE
DATA AS OF JANUARY 1, 1945

CLIMB DATA

GROSS WEIGHT LB.	BEST I.A.S. MPH (SL)	BEST I.A.S. KTS (SL)	RATE OF CLIMB F.P.M. (SL)	BEST I.A.S. MPH (5000)	BEST I.A.S. KTS (5000)	RATE OF CLIMB F.P.M. (5000)	FROM SEA LEVEL TIME MIN. (5000)	FROM SEA LEVEL FUEL USED (5000)	BEST I.A.S. MPH (10000)	BEST I.A.S. KTS (10000)	RATE OF CLIMB F.P.M. (10000)	FROM SEA LEVEL TIME MIN. (10000)	FROM SEA LEVEL FUEL USED (10000)	BEST I.A.S. MPH (15000)	BEST I.A.S. KTS (15000)	RATE OF CLIMB F.P.M. (15000)	FROM SEA LEVEL TIME MIN. (15000)	FROM SEA LEVEL FUEL USED (15000)	BEST I.A.S. MPH (20000)	BEST I.A.S. KTS (20000)	RATE OF CLIMB F.P.M. (20000)	FROM SEA LEVEL TIME MIN. (20000)	FROM SEA LEVEL FUEL USED (20000)	BEST I.A.S. MPH (25000)	BEST I.A.S. KTS (25000)	RATE OF CLIMB F.P.M. (25000)	FROM SEA LEVEL TIME MIN. (25000)	FROM SEA LEVEL FUEL USED (25000)	BEST I.A.S. MPH (30000)	BEST I.A.S. KTS (30000)	RATE OF CLIMB F.P.M. (30000)	FROM SEA LEVEL TIME MIN. (30000)	FROM SEA LEVEL FUEL USED (30000)
12,500	150	130	2600	150	130	2600	2	20	145	125	2500	4	25	145	125	2500	6	35	145	125	2400	8	45	145	125	2200	10	55	140	120	1500	13	65
13,500	150	130	2200	150	130	2200	2	20	145	125	2100	4	30	145	125	2100	7	40	145	125	1800	9	50	140	120	1800	12	60	135	115	1100	15	75
14,500	140	120	1900	140	120	1900	3	20	140	120	1800	5	35	140	120	1800	8	45	140	120	1700	11	55	140	120	1400	14	70	135	115	800	19	90

FUEL USED (U.S. GAL.) INCLUDES WARM-UP AND TAKE-OFF ALLOWANCE

LANDING DISTANCE FEET

HARD DRY SURFACE

GROSS WEIGHT LB.	BEST I.A.S. APPROACH POWER OFF MPH	BEST I.A.S. APPROACH POWER OFF KTS	AT SEA LEVEL GROUND ROLL	AT SEA LEVEL TO CLEAR 50' OBJ.	AT 3000 FEET GROUND ROLL	AT 3000 FEET TO CLEAR 50' OBJ.	AT 6000 FEET GROUND ROLL	AT 6000 FEET TO CLEAR 50' OBJ.
10,500	110	95	1180	2580	1200	2700	1420	3010
11,500	115	100	1280	2810	1390	3050	1510	3300

FIRM DRY SOD

GROSS WEIGHT LB.	AT SEA LEVEL GROUND ROLL	AT SEA LEVEL TO CLEAR 50' OBJ.	AT 3000 FEET GROUND ROLL	AT 3000 FEET TO CLEAR 50' OBJ.	AT 6000 FEET GROUND ROLL	AT 6000 FEET TO CLEAR 50' OBJ.
10,500	1310	2700	1440	2920	1560	3180
11,500	1400	2940	1530	3190	1660	3450

WET OR SLIPPERY

GROSS WEIGHT LB.	AT SEA LEVEL GROUND ROLL	AT SEA LEVEL TO CLEAR 50' OBJ.	AT 3000 FEET GROUND ROLL	AT 3000 FEET TO CLEAR 50' OBJ.	AT 6000 FEET GROUND ROLL	AT 6000 FEET TO CLEAR 50' OBJ.
10,500	2610	4600	2850	4350	3120	4720
11,500	2890	4820	3150	4810	3450	5220

NOTE: FOR GROUND TEMPERATURES ABOVE 25°C (95°F.) INCREASE APPROACH I.A.S. 10% AND ALLOW 20% INCREASE IN GROUND ROLL

OPTIMUM LANDING IS 80% OF CHART VALUES

LEGEND

I.A.S. — INDICATED AIR SPEED
M.P.H. — MILES PER HOUR
KTS. — KNOTS
F.P.M. — FEET PER MINUTE

REMARKS:

TAKE-OFF AND CLIMB DATA INCLUDE THE EFFECT OF THE DRAG INCREMENT OF EITHER ONE 150 U.S. GAL. DROP TANK OR ONE 1000 POUND BOMB IN THE 13,500 POUND CONDITION. FOR THE 14,500 POUND CONDITION, THE EFFECT OF THE DRAG INCREMENT FOR ANY COMBINATION OF TWO UNITS (TANKS, BOMBS, BOMB AND TANK) IS INCLUDED.

NOTE: TO DETERMINE FUEL CONSUMPTION IN BRITISH IMPERIAL GALLONS MULTIPLY BY 10, THEN DIVIDE BY 12

Appendix 1 of this publication shall not be carried in aircraft on combat missions or when there is a reasonable chance of its falling into the hands of the enemy.

RESTRICTED
AN-01-45HB-1
Appendix I

FLIGHT OPERATION INSTRUCTION CHART

SHEET 1 OF 3 SHEETS
CHART WEIGHT LIMITS: 12500 TO 11100 POUNDS

AIRCRAFT MODEL: F4U-4
ENGINE: R-2800-18W
EXTERNAL LOAD ITEMS: NONE

LIMITS	R.P.M.	M.P. IN. HG.	BLOWER POSITION	MIXTURE POSITION	FUEL U.S. GAL.	TIME LIMIT	CYL. TEMP.	TOTAL G.P.H.
COMBAT	2800	60	HIGH	A.R.		5 MIN.	245°C	294
	2800	60	LOW	A.R.		5 MIN.	245°C	282
	2800	60	NEUTRAL	A.R.		5 MIN.	245°C	276
MILITARY POWER	2800	54	HIGH	A.R.		5 MIN.	245°C	276
	2800	54	LOW	A.R.		5 MIN.	245°C	288
	2800	54.5	NEUTRAL	A.R.		5 MIN.	245°C	270

FOR DETAILS SEE POWER PLANT CHART, SECT III

INSTRUCTIONS FOR USING CHART: SELECT FIGURE IN FUEL COLUMN EQUAL TO OR LESS THAN AMOUNT OF FUEL TO BE USED FOR CRUISING. MOVE HORIZONTALLY TO RIGHT OR LEFT AND SELECT RANGE VALUE EQUAL TO OR GREATER THAN THE STATUTE OR NAUTICAL AIRMILES TO BE FLOWN. VERTICALLY BELOW AND OPPOSITE VALUE NEAREST DESIRED CRUISING ALTITUDE (ALT.) READ R.P.M., MANIFOLD PRESSURE (M.P.) AND MIXTURE SETTING REQUIRED.

NOTES: COLUMN I IS FOR EMERGENCY HIGH SPEED CRUISING ONLY. COLUMNS II, III, IV, AND V GIVE PROGRESSIVE INCREASE IN RANGE AT A SACRIFICE IN SPEED. AIRMILES PER GALLON (MI./GAL.) (NO WIND), GALLONS PER HR. (G.P.H.) AND TRUE AIR SPEED (T.A.S.) ARE APPROXIMATE VALUES FOR REFERENCE. RANGE VALUES ARE FOR AN AVERAGE AIRPLANE FLYING ALONE (NO WIND). TO OBTAIN BRITISH IMPERIAL GAL. (OR G.P.H.) MULTIPLY U.S. GAL. (OR G.P.H.) BY 10, THEN DIVIDE BY 12.

COLUMN I — Range in Airmiles

STATUTE AT S.L.	NAUTICAL AT S.L.	APPROX. M.P.H.	T.A.S. KTS.	TOT. G.P.H.	MIX. TURE	M.P. IN-CHES	R.P.M.	FUEL U.S. GAL.	PRESS. ALT. FEET
330	290	421	366		A.L.	F.T.	2800	233	30000
300	260	398	346		A.L.	F.T.	2800	210	25000
270	235	387	336		A.L.	48	2800	190	20000
240	205	367	316		A.L.	48	2800	170	15000
205	180	—	—			F.T.	2600	130	10000
175	150	—	—				2600	110	5000
140	120	—	—				2600	90	
110	95	325	300		A.L.	F.T.	2600	70	10000
80	70	315	290		A.L.	43.5	2600	50	5000
50	45	313	272		A.L.	43.5	2600	30	S.L.

COLUMN II — Range in Airmiles (SUBTRACT FUEL ALLOWANCES NOT AVAILABLE FOR CRUISING)

STATUTE	NAUTICAL	R.P.M.	M.P. IN-CHES	MIX. TURE	TOT. G.P.H.	APPROX. M.P.H. T.A.S. KTS.	FUEL U.S. GAL.
545	475	2200	40	A.L.	153	397 344	
495	430	2400	40	A.L.	147	382 331	
440	385	2400	39	A.L.	135	350 304	
390	340	2400	38	A.L.	124	323 280	
340	290						
285	250						
235	205						
180	160	2400	32	A.L.	121	315 273	
130	115	2300	32	A.L.	113	294 255	
80	70	2300	32	A.L.	106	275 239	

COLUMN III — Range in Airmiles

STATUTE	NAUTICAL	R.P.M.	M.P. IN-CHES	MIX. TURE	TOT. G.P.H.	APPROX. M.P.H. T.A.S. KTS.
710	620					
645	560					
575	500					
510	440					
440	380					
375	325					
305	265					
235	205	2100	30	A.L.	89	300 260
170	145	2100	31	A.L.	85	286 248
100	90					

COLUMN IV — Range in Airmiles

STATUTE	NAUTICAL	R.P.M.	M.P. IN-CHES	MIX. TURE	TOT. G.P.H.	APPROX. M.P.H. T.A.S. KTS.
830	720					
750	650					
670	580					
590	515					
515	445					
435	375					
355	310					
275	240	1800	29	A.L.	63	249 216
195	170	1600	28	A.L.	63	250 217
120	100	1600	29	A.L.	60	235 204
		1600	30	A.L.	56	220 191

COLUMN V — Maximum Air Range

STATUTE	NAUTICAL	FUEL U.S. GAL.	PRESS. ALT. FEET	R.P.M.	M.P. IN-CHES	MIX. TURE	TOT. G.P.H.	APPROX. M.P.H. T.A.S. KTS.
	840	233	30000					
965	755	210	25000					
870	675	190	20000					
780	595	170						
690	520	130	15000					
600	440	110						
505	340	90						
415	285	70	10000					
325	200	50	5000					
230	120	30	S.L.	1400	26	A.L.	42	191 166
140				1300	27	A.L.	39	180 156

MAXIMUM CONTINUOUS

BLOWER: H / L / N

SPECIAL NOTES

(1) MAKE ALLOWANCE FOR WARM-UP, TAKE-OFF, AND CLIMB (SEE THE TAKE-OFF, CLIMB AND LANDING CHART) PLUS ALLOWANCE FOR WIND, RESERVE AND COMBAT AS REQUIRED.

DATA AS OF JANUARY 1, 1945

EXAMPLE

AT 12000 LB. GROSS WEIGHT WITH 308 GAL. OF FUEL (AFTER DEDUCTING TOTAL ALLOWANCE OF 25 GAL. FOR WARM-UP, TAKE-OFF AND CLIMB) TO FLY 540 STAT. AIRMILES AT 10000 FT. ALTITUDE, MAINTAIN 2000 R.P.M., AUTO LEAN, AIRMILES AT 10000 FT. MANIFOLD PRESSURE WITH MIXTURE SET: AUTO LEAN.

LEGEND

ALT.: PRESSURE ALTITUDE
M.P.: MANIFOLD PRESSURE
G.P.H.: U.S. GAL. PER HOUR
T.A.S.: TRUE AIR SPEED
KTS.: KNOTS

S.L.: SEA LEVEL
A.R.: AUTO RICH
A.L.: AUTO LEAN
F.T.: FULL THROTTLE

RED FIGURES ARE PRELIMINARY DATA, SUBJECT TO REVISION AFTER FLIGHT CHECK.

Appendix 1 of this publication shall not be carried in aircraft on combat missions or when there is a reasonable chance of its falling into the hands of the enemy.

FLIGHT OPERATION INSTRUCTION CHART

Appendix 1 — **RESTRICTED** — AN-01-45HB-1

SHEET 2 OF 3 SHEETS

CHART WEIGHT LIMITS: 13500 TO 11200 POUNDS

AIRCRAFT MODEL: F4U-4
ENGINE: R-2800-18W

EXTERNAL LOAD ITEMS: ONE 150 GAL. DROP TANK OR ONE 1000 POUND BOMB.

LIMITS	R.P.M.	M.P. IN. HG.	BLOWER POSITION	MIXTURE POSITION	TIME LIMIT	CYL. TEMP.	TOTAL G.P.H.
COMBAT	2800	60	HIGH	A.R.	5 MIN.	245°C	294
	2800	60	LOW				282
	2800	60	NEUTRAL				276
MILITARY POWER	2800	54	HIGH	A.R.	5 MIN.	245°C	276
	2800	54	LOW				288
	2800	54.5	NEUTRAL				270

SECT III POWER PLANT CHART FOR DETAILS SEE

NOTES: COLUMN I IS FOR EMERGENCY HIGH SPEED CRUISING ONLY. COLUMNS II, III, IV, AND V GIVE PROGRESSIVE INCREASE IN RANGE AT A SACRIFICE IN SPEED. AIRMILES PER GALLON (MI./GAL.) (NO WIND), GALLONS PER HR. (G.P.H.) AND TRUE AIR SPEED (T.A.S.) ARE APPROXIMATE VALUES FOR REFERENCE. RANGE VALUES ARE FOR AN AVERAGE AIRPLANE FLYING ALONE (NO WIND). TO OBTAIN BRITISH IMPERIAL GAL. (OR G.P.H.) MULTIPLY U.S. GAL. (OR G.P.H.) BY 10, THEN DIVIDE BY 12.

INSTRUCTIONS FOR USING CHART: SELECT FIGURE IN FUEL COLUMN EQUAL TO OR LESS THAN AMOUNT OF FUEL TO BE USED FOR CRUISING. MOVE HORIZONTALLY TO RIGHT OR LEFT AND SELECT RANGE VALUE EQUAL TO OR GREATER THAN TRUE STATUTE OR NAUTICAL AIRMILES TO BE FLOWN. VERTICALLY BELOW AND OPPOSITE VALUE NEAREST DESIRED CRUISING ALTITUDE (ALT.) READ R.P.M., MANIFOLD PRESSURE (M.P.) AND MIXTURE SETTING REQUIRED.

RANGE IN AIRMILES

COLUMN I — (2.3 NAUT.) MI./GAL

NAUTICAL AT S.L.	STATUTE	FUEL U.S. GAL.
450	520	383 / 350
410	470	315
360	420	280
320	370	245
270	315	210
230	260	175
180	210	140
140	160	105
90	105	70
45	50	35

COLUMN II — (2.6 STAT. / 2.3 NAUT.) MI./GAL

SUBTRACT FUEL ALLOWANCES

STATUTE	NAUTICAL
910	790
820	710
730	630
640	550
550	470
455	395
360	320
270	240
180	160
90	80

COLUMN III — (3.0 STAT. / 2.6 NAUT.) MI./GAL

ALLOWANCES NOT AVAILABLE FOR CRUISING(1)

STATUTE	NAUTICAL
1045	905
940	815
835	725
730	635
625	545
520	450
415	360
315	270
210	180
105	90

COLUMN IV — (3.4 STAT. / 3.0 NAUT.) MI./GAL

STATUTE	NAUTICAL
1190	1040
1070	930
950	825
835	720
715	620
595	515
475	415
355	310
240	205
120	105

COLUMN V — MAXIMUM AIR RANGE

STATUTE	NAUTICAL	FUEL U.S. GAL.
1375	1190	383 / 350
1240	1075	315
1100	955	280
965	835	245
825	715	210
690	595	175
550	480	140
415	360	105
275	240	70
140	120	35

MAXIMUM CONTINUOUS

PRESS. ALT. FEET	R.P.M.	M.P. IN. CHES	MIX. TURE	APPROX. G.P.H.	TOT. M.P.H.	T.A.S. KTS.
30000						
25000	2400	40	A.R.	139 / 214	362 / 393	314 / 341
20000	2400	38	A.L.	132 / 201	344 / 373	298 / 324
15000	2400	36	A.L.	122 / 217	316 / 363	274 / 315
10000	2200	34	A.L.	110 / 214	287 / 345	249 / 300
5000	2300	31	A.L.	111 / 193	288 / 324	250 / 281
S.L.	2300	32	A.L.	102 / 206	266 / 316	231 / 274
	2200	32	A.L.	97 / 198	253 / 295	220 / 256

(pressure altitudes continued)

PRESS. ALT. FEET	R.P.M.	M.P. IN. CHES	MIX. TURE	APPROX. G.P.H.	T.A.S. M.P.H.	KTS.
30000						
25000	2200	31	A.L.	93	278	242
20000	2100	32	A.L.	89	266	231
15000	1800	32	A.L.	78	232	202
10000	1800	28	A.L.	67	228	198
5000	1600	31	A.L.	62	211	183
S.L.	1500	26	A.L.	46	183	159
	1400	28	A.L.	45	178	154

BLOWER: H, L, N

SPECIAL NOTES

(1) MAKE ALLOWANCE FOR WARM-UP, TAKE-OFF, AND CLIMB (SEE THE TAKE-OFF, CLIMB AND LANDING CHART) PLUS ALLOWANCE FOR WIND, RESERVE AND COMBAT AS REQUIRED.

DATA AS OF JANUARY 1, 1945

EXAMPLE

AT 12000 LB. GROSS WEIGHT WITH 353 GAL. OF FUEL (AFTER DEDUCTING TOTAL ALLOWANCE OF 30 GAL. FOR WARM-UP, TAKE-OFF AND CLIMB) TO FLY 1025 STAT. AIRMILES AT 10000 FT. ALTITUDE. MAINTAIN 2200 R.P.M. AND 27 IN. MANIFOLD PRESSURE WITH MIXTURE SET, AUTO LEAN.

LEGEND

ALT.: PRESSURE ALTITUDE
M.P.: MANIFOLD PRESSURE
G.P.H.: U.S. GAL. PER HOUR
T.A.S.: TRUE AIR SPEED
KTS.: KNOTS
S.L.: SEA LEVEL
A.R.: AUTO RICH
A.L.: AUTO LEAN
F.T.: FULL THROTTLE

RED FIGURES ARE PRELIMINARY DATA, SUBJECT TO REVISION AFTER FLIGHT CHECK.

Appendix 1 of this publication shall not be carried in aircraft on combat missions or when there is a reasonable chance of its falling into the hands of the enemy.

RESTRICTED
AN-01-45HB-1

Appendix 1

FLIGHT OPERATION INSTRUCTION CHART
SHEET 3 OF 3 SHEETS
CHART WEIGHT LIMITS: 14500 TO 12200 POUNDS

AIRCRAFT MODEL F4U-4
ENGINE: R-2800-18W

EXTERNAL LOAD ITEMS
ONE 150 GAL. DROP TANK AND ONE 1000 LB. BOMB (OR TWO 1000 LB. BOMBS)

LIMITS	R.P.M.	M.P. IN. HG.	BLOWER POSITION	MIXTURE POSITION	TIME LIMIT	CYL. TEMP.	TOTAL G.P.H.
COMBAT	2800	60 / 60 / 60	HIGH / LOW / NEUTRAL	A.R.	5 MIN.	245°C	294 / 282 / 276
MILITARY POWER	2800	54 / 54 / 54.5	HIGH / LOW / NEUTRAL	A.R.	5 MIN.	245°C	276 / 288 / 270

FOR DETAILS SEE POWER PLANT CHART, SECT. III.

NOTES: COLUMN I IS FOR EMERGENCY HIGH SPEED CRUISING ONLY. COLUMNS II, III, IV, AND V GIVE PROGRESSIVE INCREASE IN RANGE AT A SACRIFICE IN SPEED. AIRMILES PER GALLON (MI./GAL.) (NO WIND), GALLONS PER HR. (G.P.H.) AND TRUE AIR SPEED (T.A.S.) ARE APPROXIMATE VALUES FOR REFERENCE. RANGE VALUES ARE FOR AN AVERAGE AIRPLANE FLYING ALONE (NO WIND). TO OBTAIN BRITISH IMPERIAL GAL. (OR G.P.H.) MULTIPLY U.S. GAL. (OR G.P.H.) BY 10, THEN DIVIDE BY 12.

INSTRUCTIONS FOR USING CHART: SELECT FIGURE IN FUEL COLUMN EQUAL TO OR LESS THAN AMOUNT OF FUEL TO BE USED FOR CRUISING. MOVE HORIZONTALLY TO RIGHT OR LEFT AND SELECT RANGE VALUE EQUAL TO OR GREATER THAN THE STATUTE OR NAUTICAL AIRMILES TO BE FLOWN. VERTICALLY BELOW AND OPPOSITE VALUE NEAREST DESIRED CRUISING ALTITUDE (ALT.) READ R.P.M., MANIFOLD PRESSURE (M.P.) AND MIXTURE SETTING REQUIRED.

COLUMN I

RANGE IN AIRMILES		APPROX.			M.P. IN-CHES	MIX-TURE	FUEL U.S. GAL.
STATUTE AT S.L.	NAUTICAL AT S.L.	TOT. G.P.H.	T.A.S. M.P.H.	KTS.			
495	430	214	367	318	F.T.	A.L.	383 / 350
445 / 395 / 345	385 / 340 / 300	201 / 217 / 214	350 / 342 / 324	304 / 297 / 282	48 / 48 / 48	A.L. / A.L. / A.L.	315 / 280 / 245
295 / 250 / 195	255 / 215 / 170	193 / 206 / 198	305 / 298 / 279	265 / 258 / 242	F.T. / 43.5 / 43.5	A.L. / A.L. / A.L.	210 / 175 / 140
150 / 100 / 50	130 / 85 / 45						105 / 70 / 35

MAXIMUM CONTINUOUS — R.P.M. 2600 (H), 2600 (L), 2600 (N)

COLUMN II
SUBTRACT FUEL ALLOWANCES NOT AVAILABLE FOR CRUISING(1)

RANGE IN AIRMILES		(0.1 NAUT.) MI./GAL.					
STATUTE	NAUTICAL	R.P.M.	M.P. IN-CHES	MIX-TURE	TOT. G.P.H.	T.A.S. M.P.H.	KTS.
840	730	2400	39	A.L.	137	329	285
735 / 670 / 590	655 / 580 / 510	2400 / 2400 / 2300	38 / 36 / 33	A.L. / A.L. / A.L.	132 / 122 / 110	318 / 293 / 263	276 / 254 / 228
505 / 420 / 335	435 / 365 / 290	2300 / 2300 / 2200	31 / 31 / 32	A.L. / A.L. / A.L.	112 / 104 / 99	269 / 250 / 237	233 / 217 / 206
230 / 170 / 85	220 / 145 / 75						

COLUMN III
RANGE IN AIRMILES		(0.2 STAT. (0.1 NAUT.) MI./GAL.)					
STATUTE	NAUTICAL	R.P.M.	M.P. IN-CHES	MIX-TURE	TOT. G.P.H.	T.A.S. M.P.H.	KTS.
950	825	2100	32	A.L.	94	255	221
855 / 760 / 670	745 / 670 / 580						
570 / 475 / 380	495 / 415 / 330	1800	29	A.L.	80	218	189
285 / 190 / 95	250 / 165 / 80						

COLUMN IV
RANGE IN AIRMILES		(0.3 STAT. (0.2 NAUT.) MI./GAL.)					
STATUTE	NAUTICAL	R.P.M.	M.P. IN-CHES	MIX-TURE	TOT. G.P.H.	T.A.S. M.P.H.	KTS.
1070	930						
965 / 860 / 750	835 / 745 / 650	1800	29	A.L.	67	206	179
645 / 535 / 430	560 / 465 / 370						
320 / 215 / 105	280 / 185 / 90	1700 / 1700	30 / 31	A.L. / A.L.	70 / 65	214 / 198	186 / 172

FUEL U.S. GAL.	PRESS. ALT. FEET
383 / 350	30000
315 / 280 / 245	25000 / 20000 / 15000
210 / 175 / 140	10000 / 5000 / S.L.
105 / 70 / 35	

COLUMN V
RANGE IN AIRMILES		
STATUTE	NAUTICAL	
1185	1030	
1065 / 950 / 830	925 / 825 / 720	
710 / 590 / 475	615 / 515 / 410	
355 / 235 / 120	310 / 205 / 105	

MAXIMUM AIR RANGE

PRESS. ALT. FEET	R.P.M.	M.P. IN-CHES	MIX-TURE	APPROX. TOT. G.P.H.	T.A.S. M.P.H.	KTS.
30000						
25000 / 20000 / 15000						
10000 / 5000 / S.L.	1500 / 1400	26 / 29	A.L. / A.L. / A.L.	50 / 51	170 / 171	148 / 149

SPECIAL NOTES
(1) MAKE ALLOWANCE FOR WARM-UP, TAKE-OFF AND CLIMB (SEE THE TAKE-OFF, CLIMB AND LANDING CHART) PLUS ALLOWANCE FOR WIND, RESERVE AND COMBAT AS REQUIRED.

DATA AS OF: JANUARY 1, 1945

EXAMPLE
AT 14000 LB. GROSS WEIGHT WITH 338 GAL. OF FUEL (AFTER DEDUCTING TOTAL ALLOWANCE OF 45 GAL. FOR WARM-UP, TAKE-OFF AND CLIMB) TO FLY 950 STAT. AIRMILES AT 15000 FT. ALTITUDE, MAINTAIN 1800 R.P.M. AND 29 IN. MANIFOLD PRESSURE WITH MIXTURE SET: AUTO LEAN.

LEGEND
ALT.: PRESSURE ALTITUDE
M.P.: MANIFOLD PRESSURE
G.P.H.: U.S. GAL. PER HOUR
T.A.S.: TRUE AIR SPEED
KTS.: KNOTS

S.L.: SEA LEVEL
A.R.: AUTO RICH
A.L.: AUTO LEAN
F.T.: FULL THROTTLE

RED FIGURES ARE PRELIMINARY DATA, SUBJECT TO REVISION AFTER FLIGHT CHECK.

Appendix 1 of this publication shall not be carried in aircraft on combat missions or when there is a reasonable chance of its falling into the hands of the enemy.

RESTRICTED

Appendix 1

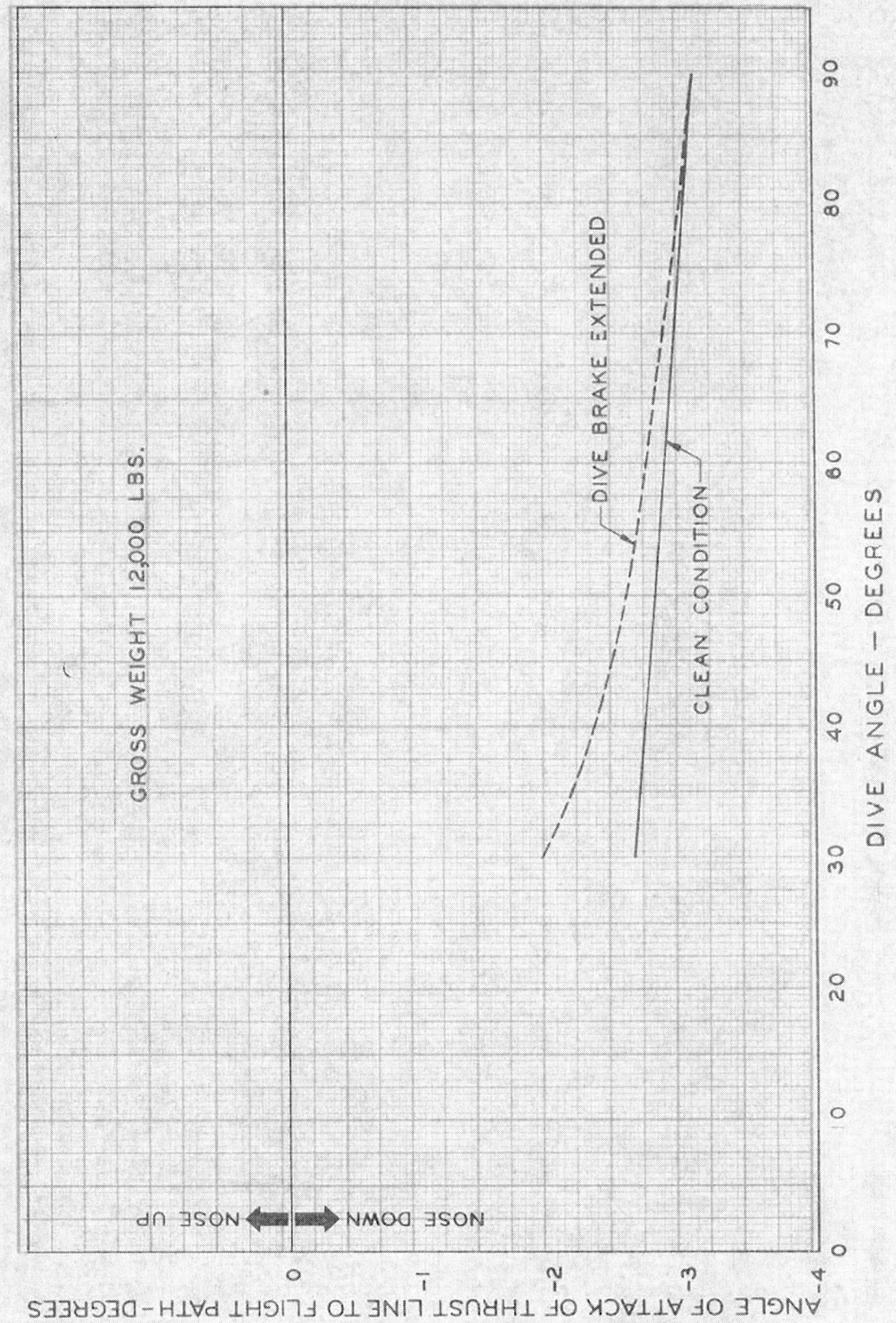

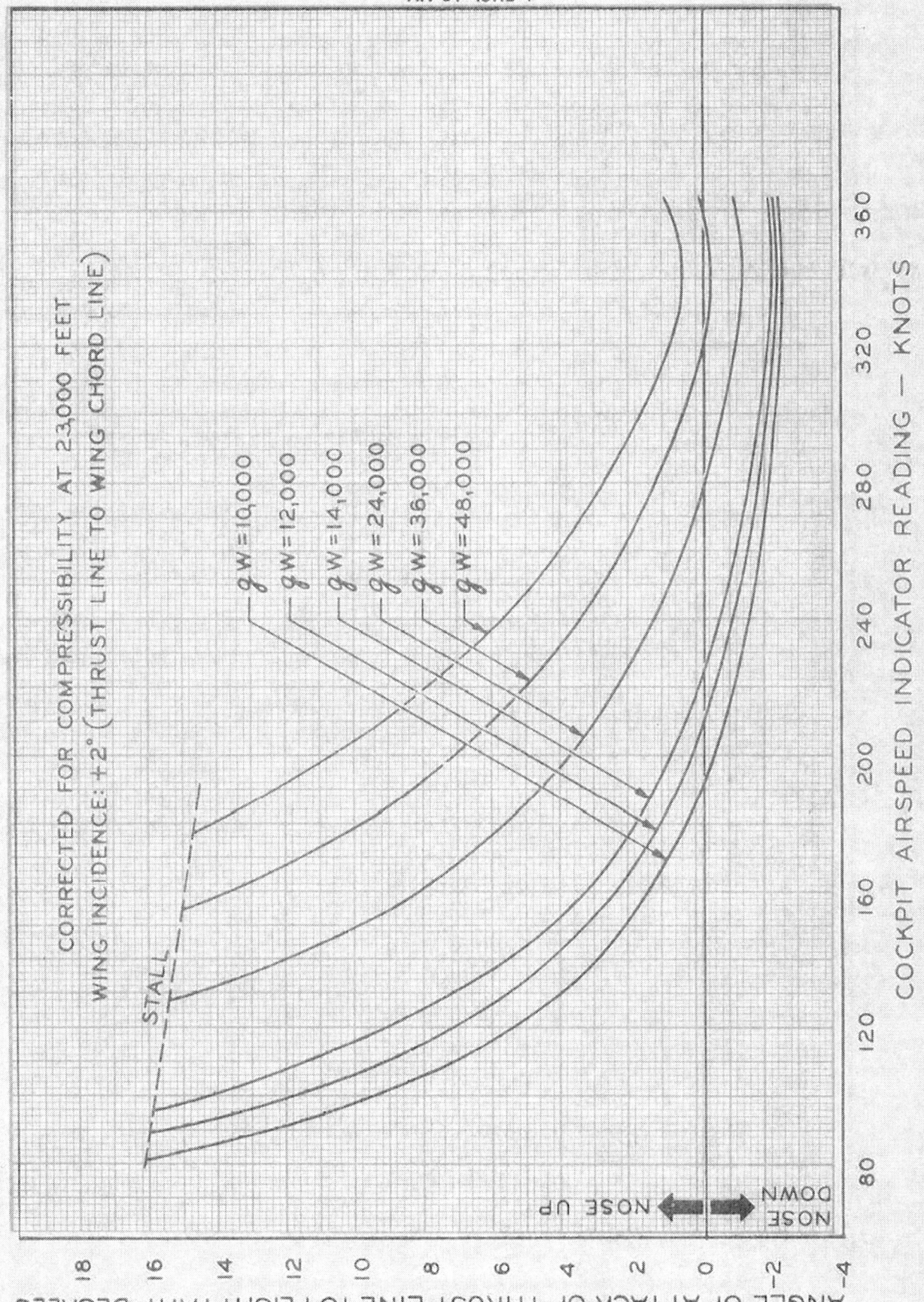

1. ENGINE CALIBRATION CURVE.

a. EXAMPLES OF USE.

(1) This curve can be used to set operating conditions or to determine engine power at any operating condition within the recommended operating limits of the engine. The curve to the left is for neutral blower operation, and the curve in the center for low blower operation, and the curve to the right for high blower operation. Part throttle conditions are those to the left of the oblique, heavy, dashed line in all blower sections; full throttle conditions are those to the right of these lines. Use auto rich mixtures at take-off, military power, and combat power operations. Use auto lean mixtures at lower powers. Cylinder head temperature limits must be observed, however, so use auto rich when cooling is inadequate in auto lean.

b. HIGH POWER—Rich mixture in Military and Combat Power, lean mixture in Normal Rated Power and below. (Part Throttle).

(1) When high power climb is desired, operate along one of the constant manifold pressure —RPM lines (sloping lines labeled with manifold pressure and RPM). For rated power climb, use 2600 RPM, 43.5 inches Hg. in neutral blower from sea level to critical altitude. Above critical, shift to low blower, 48 inches Hg., when the manifold pressure has decreased to 40 inches Hg. Above low blower critical altitude, shift to high blower, 47.5 inches Hg., when the manifold pressure has dropped to 45 inches Hg.

(2) Select the desired level flight condition from a point on one of the designated lines, or, if an intermediate condition is desired, any manifold pressure-RPM combination represented in the full throttle portions of the chart can be used for part-throttle operation.

c. CRUISING POWER—LEAN MIXTURE (PART THROTTLE).

(1) For power conditions below 1250 BHP in neutral blower, 1200 BHP in low blower, and 1175 BHP in high blower, the maximum recommended manifold pressures are essentially independent of RPM.

d. TO DETERMINE HORSEPOWER — ANY POWER CONDITION.

(1) Knowing RPM and manifold pressure, spot the condition in the full-throttle portion of the section of the chart for the blower ratio in which he engine is operating.

(2) Draw a line through the point determined, parallel to the constant manifold pressure — RPM lines shown. Read horsepower at the intersection of this line with the observed pressure altitude.

e. PRESSURE ALTITUDE.

(1) Determine the amount the barometric pressure (altimeter window reading) is above or below 29.92 inches Hg.

(2) Add 100 feet for each 0.1 inch Hg. below 29.92; subtract 100 feet for each 0.1 inch Hg. above 29.92.

Appendix 1 of this publication shall not be carried in aircraft on combat missions or when there is a reasonable chance of its falling into the hands of the enemy.

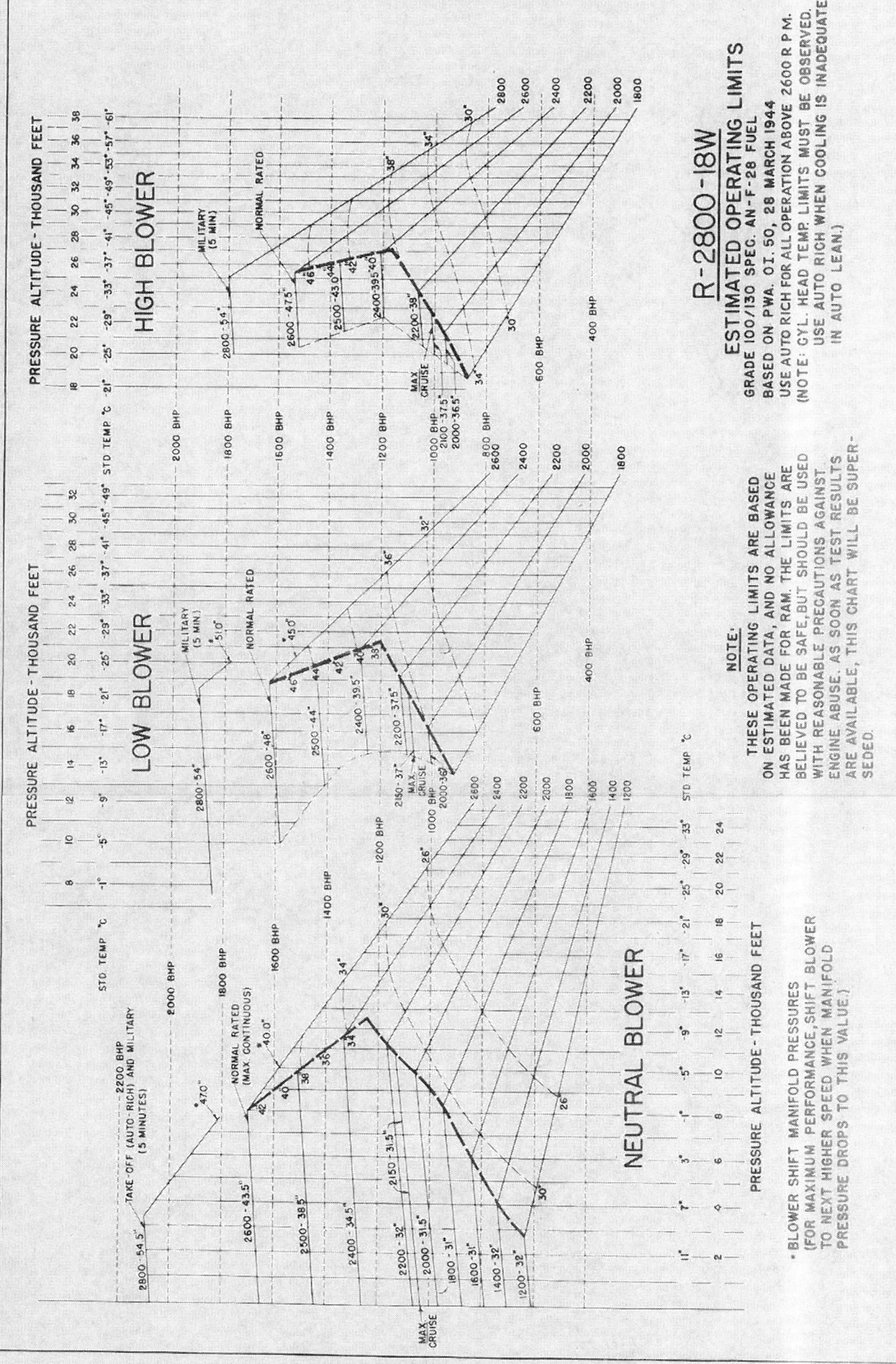

F4U

Illustrated ASSEMBLY BREAKDOWN

This booklet has been prepared by the Publications Section of the Engineering Department at the request of Tool Engineering. It is designed to be used as a ready reference for major assembly members of the airplane by all departments of our Company and its subcontractors.

SEPTEMBER 1, 1944

M. SCHWARTZ

CHANCE VOUGHT AIRCRAFT • STRATFORD, CONNECTICUT

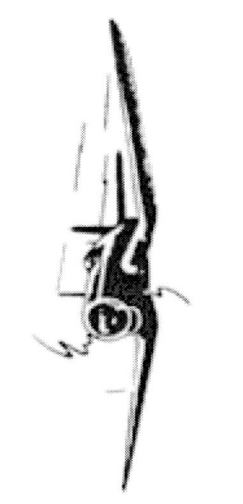

INDEX

	Page
PROFILE	2
MAJOR ASSEMBLIES	3
POWER PLANT	4, 5
FRONT SECTION	6, 7
MID SECTION	8, 9
AFT SECTION	10, 11
CENTER SECTION	12, 13
OUTER PANEL	14
WING ERECTION	15
TAIL ERECTION	16

This document contains information affecting the national defense of the United States within the meaning of the Espionage Act, 50; U.S.C. 31 and 32, as amended. Its transmission or the revelation of its contents in any manner to an unauthorized person is prohibited by law.

PROFILE

General Arrangement	VS-40601
Armor Plating Arrange	VS-44014
Bomb & Drop Tank Instal	VS-33540
Box of Loose Equipment	VS-45075
Catapult Arrangement	VS-37481
Droppable Tank Assem — Aux	VS-40303
Electrical System	VS-44018
Electrical System Wiring Diagram	VS-44019
Elevator, Rudder & Tab Controls	VS-40404
Forward Fuselage Unit	VS-40250
Front & Center Sect Unit Assem	VS-40225
Fuselage Assem Body Group	VS-40200
Ground Handling Equipment	VS-45165
Hydraulic System Diagram	VS-42998
Hydraulic System Rigging	VS-42997
Inboard Profile	VS-40600
Lines Drawing Body	VS-40201
Lubricating System	VS-40307
Lubricating System Diagram	VS-43064
Markings F4U-4 Airplane	VS-40900
Oiling Diagram	VS-43065
Parachute & Liferaft Arrange	VS-45005
Radio & Cable Instal Identif	VS-40535
Radio & Cable Instal Commun	VS-40525
Rear Fuselage Unit	VS-40210
Tail Rise Calculation	VS-44021
Towing & Tie-Down Diagram	VS-45003
Surface Controls Diagram	VS-44017

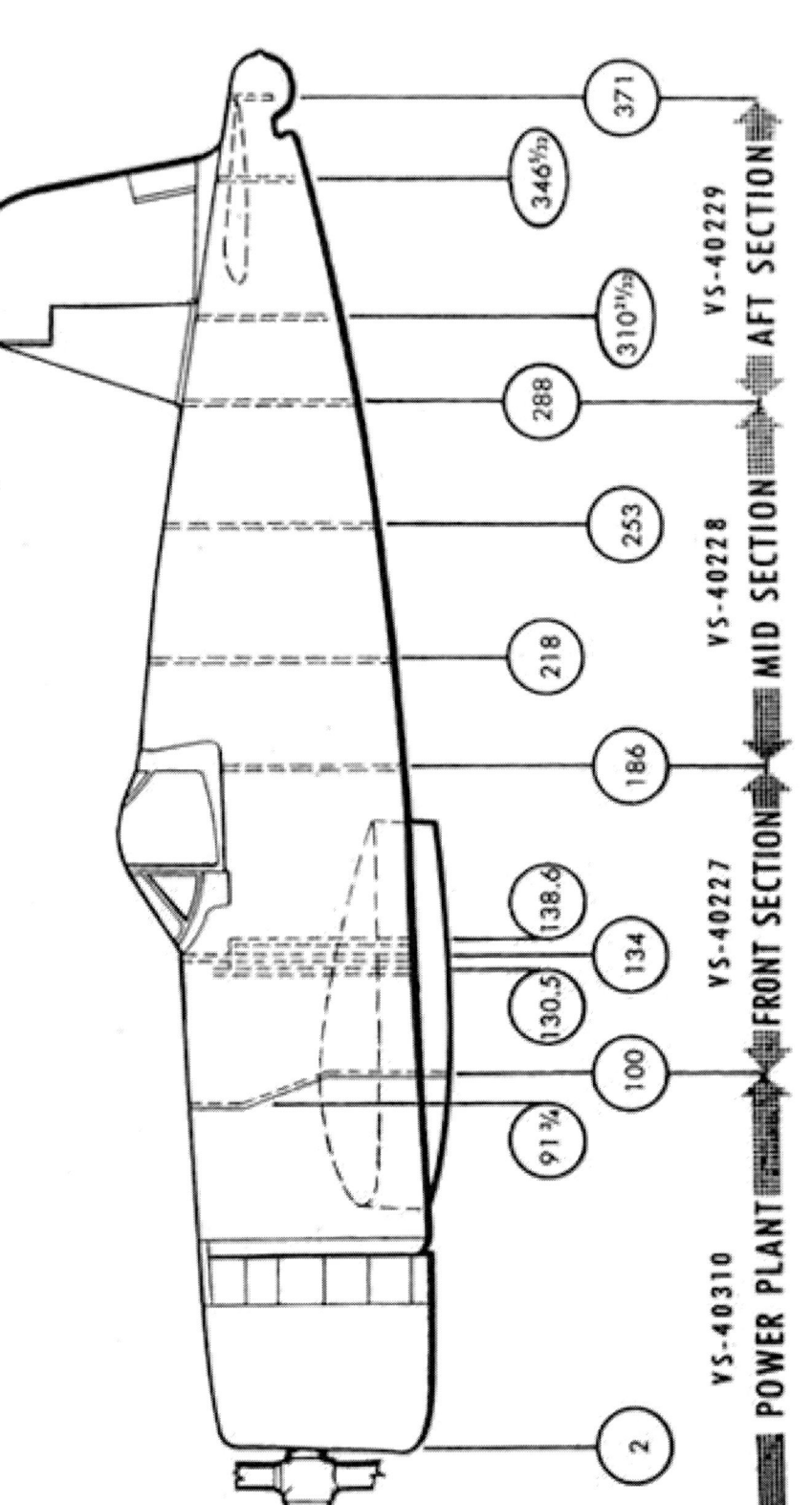

MAJOR ASSEMBLIES

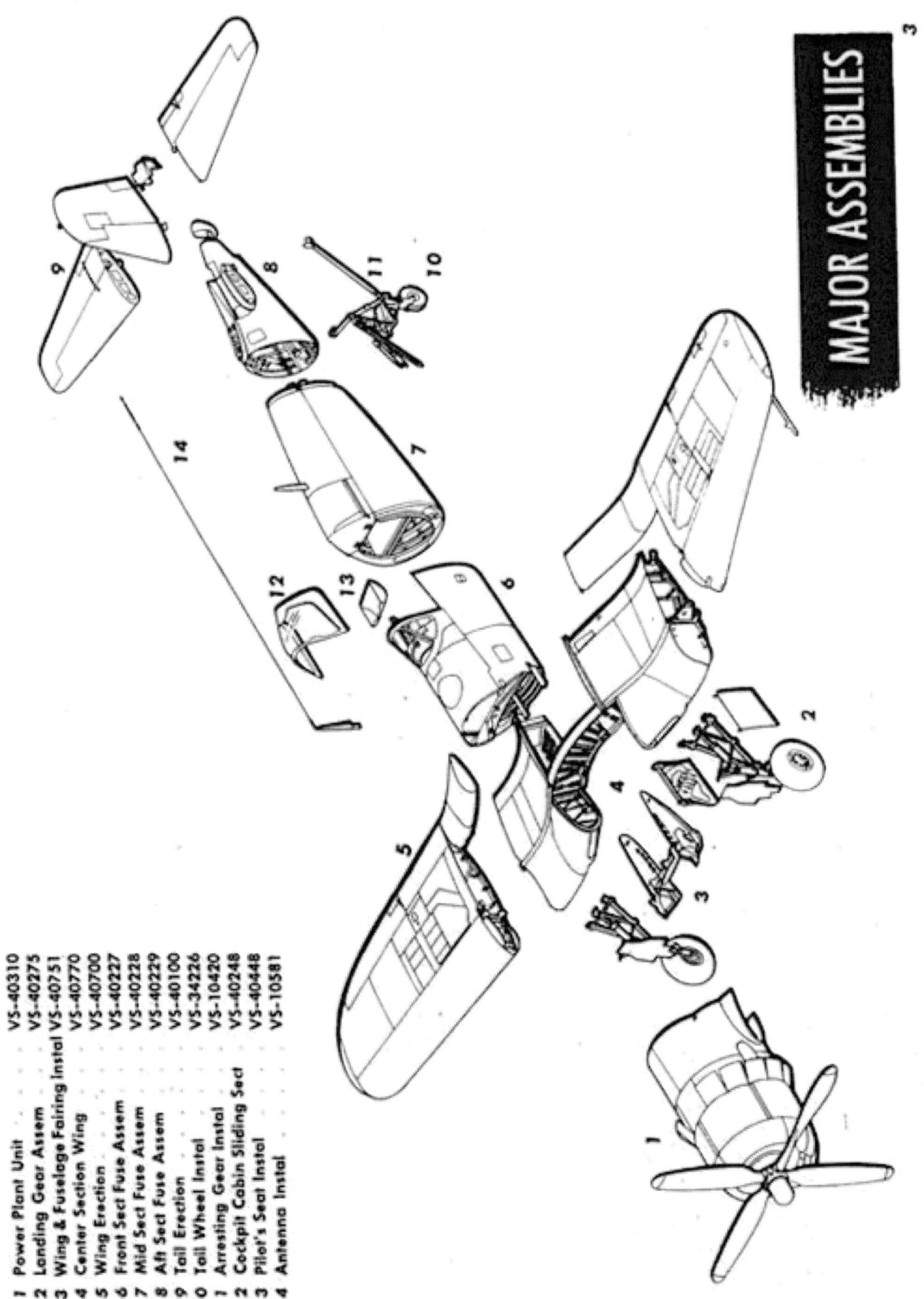

1	Power Plant Unit	VS-40310
2	Landing Gear Assem	VS-40275
3	Wing & Fuselage Fairing Instal	VS-40751
4	Center Section Wing	VS-40770
5	Wing Erection	VS-40700
6	Front Sect Fuse Assem	VS-40227
7	Mid Sect Fuse Assem	VS-40228
8	Aft Sect Fuse Assem	VS-40229
9	Tail Erection	VS-40100
10	Tail Wheel Instal	VS-34226
11	Arresting Gear Instal	VS-10420
12	Cockpit Cabin Sliding Sect	VS-40248
13	Pilot's Seat Instal	VS-40448
14	Antenna Instal	VS-10581

POWER PLANT

Power Plant Unit	VS-40310
Oil Tank Assem	VS-43306
Hydraulic Accumulator	VS-12281
Hydraulic Reservoir Assem	VS-48088
Engine Mount & Attach Parts	VS-40640
Carburetor Air Box Assem	VS-40335
Automatic Cowl Flap Controls	VS-
Intermediate Intercool Duct Assem	VS-40340
Magneto	DF-18LU-2
Distributor	
Engine—Pratt & Whitney	R-2800-18-W
Propeller Blade—Hamilton Std	6507-A-0
Propeller Hub—Hamilton Std	24E60
Aux Stage Air Exit Duct Assem	VS-40337
Blower Drain Instal	VS-40308
Electrical System Instal	VS-40430
Engine Accessories	VS-40302
Engine Controls Instal	VS-40301-1
Fuel System Instal	VS-40325-1
Hydraulic System Instal	VS-40271
Instrument Instal	VS-40470-1
Lubricating System	VS-40307-1
Power Plant Instal	VS-40300
Removable Section	VS-40350
Res, Valves, Accum, & Pan Assem	VS-48001
Starter Instal—Jack & Heintz	JH4NER
Intercool Air Ent Duct Assem—L H	VS-40338
Intercool Air Ent Duct Assem—R H	VS-40348
Intercool Air Exit Duct Assem	VS-40339
Aux Stage Air Entrance Duct Assem	VS-40336
Intercooler—Engine Access	VS-40395
Intercooler Suppl Rib Assem—L H	VS-40323
Intercooler Suppl Rib Assem—R H	VS-40324
Vac Pump & Oil Separator Instal	VS-40396
Water Injection System	VS-40327
Water Tank Assem—Water Inject	VS-40328

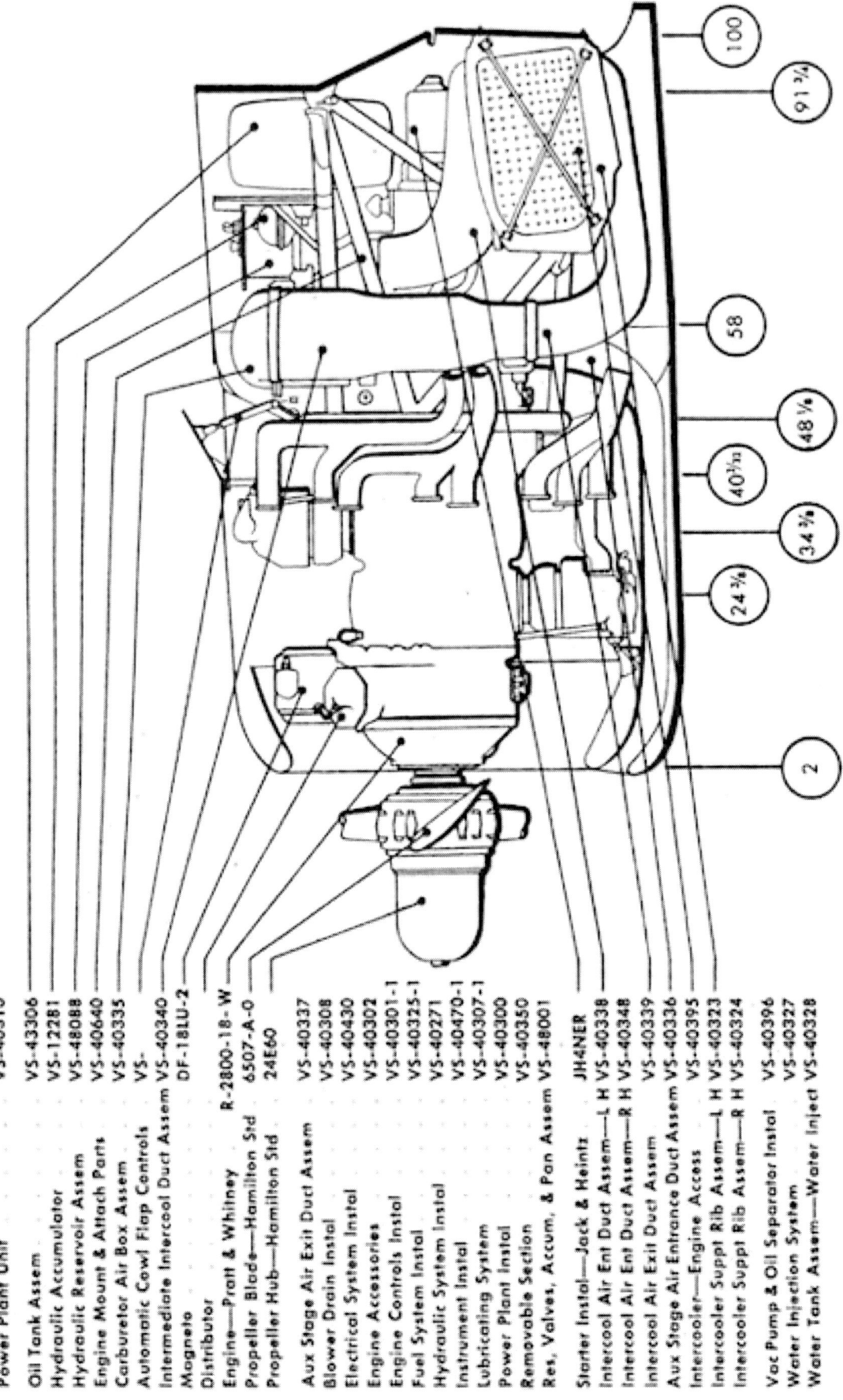

POWER PLANT

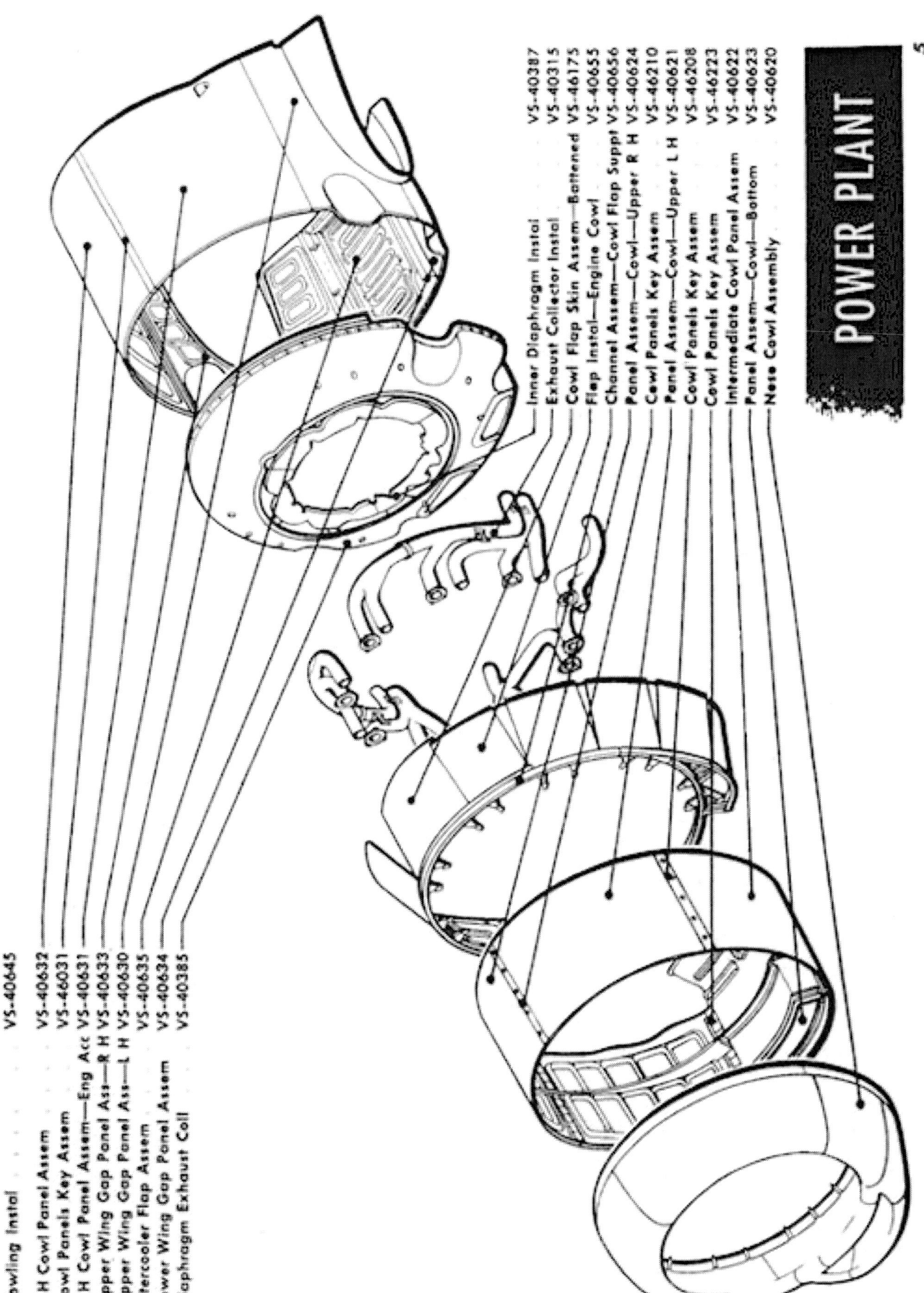

Cowling Instal	VS-40645
R H Cowl Panel Assem	VS-40632
Cowl Panels Key Assem	VS-46031
L H Cowl Panel Assem—Eng Acc	VS-40631
Upper Wing Gap Panel Ass—R H	VS-40633
Upper Wing Gap Panel Ass—L H	VS-40630
Intercooler Flap Assem	VS-40635
Lower Wing Gap Panel Assem	VS-40634
Diaphragm Exhaust Coll	VS-40385
Inner Diaphragm Instal	VS-40387
Exhaust Collector Instal	VS-40315
Cowl Flap Skin Assem—Battened	VS-46175
Flap Instal—Engine Cowl	VS-40655
Channel Assem—Cowl Flap Suppt	VS-40656
Panel Assem—Cowl Flap—Upper R H	VS-40624
Cowl Panels Key Assem	VS-46210
Panel Assem—Cowl—Upper L H	VS-40621
Cowl Panels Key Assem	VS-46208
Cowl Panels Key Assem	VS-46223
Intermediate Cowl Panel Assem	VS-40622
Panel Assem—Cowl—Bottom	VS-40623
Nose Cowl Assembly	VS-40620

FRONT SECTION

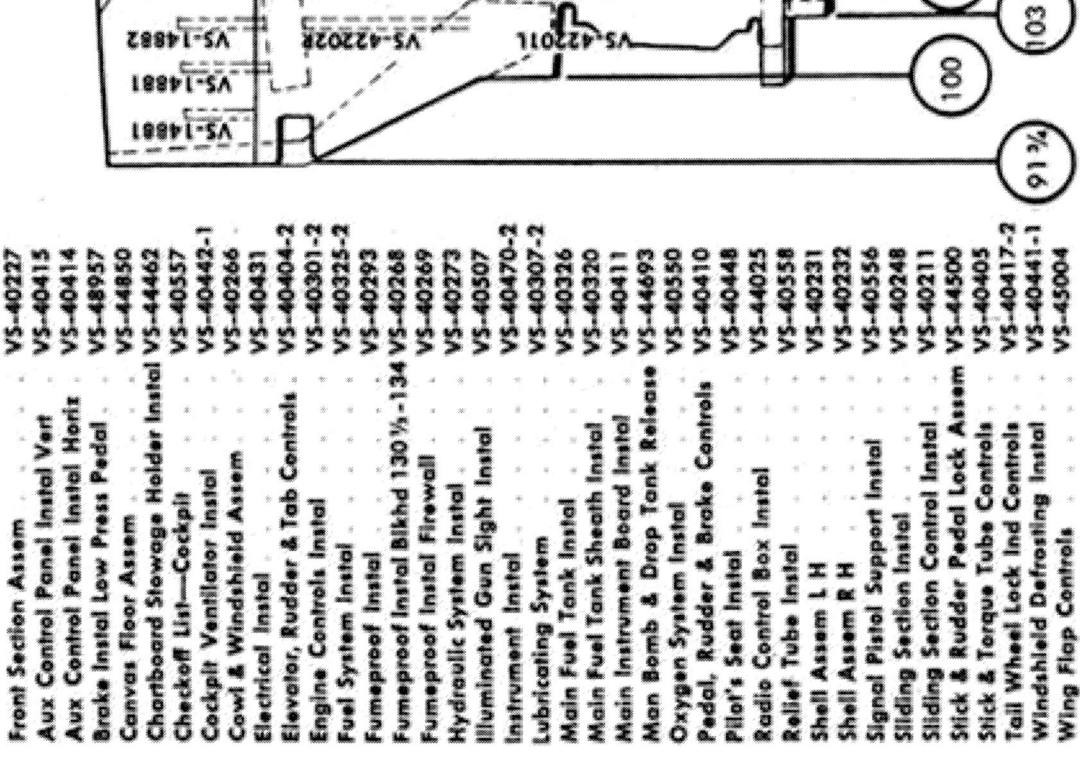

Front Section Assem	VS-40227
Aux Control Panel Instal Vert	VS-40415
Aux Control Panel Instal Horiz	VS-40414
Brake Instal Low Press Pedal	VS-48957
Canvas Floor Assem	VS-44850
Chartboard Stowage Holder Instal	VS-44462
Checkoff List—Cockpit	VS-40557
Cockpit Ventilator Instal	VS-40442-1
Cowl & Windshield Assem	VS-40266
Electrical Instal	VS-40431
Elevator, Rudder & Tab Controls	VS-40404-2
Engine Controls Instal	VS-40301-2
Fuel System Instal	VS-40325-2
Fumeproof Instal	VS-40293
Fumeproof Instal Blkhd 130 ½ –134	VS-40268
Fumeproof Instal Firewall	VS-40269
Hydraulic System Instal	VS-40273
Illuminated Gun Sight Instal	VS-40507
Instrument Instal	VS-40470-2
Lubricating System	VS-40307-2
Main Fuel Tank Instal	VS-40326
Main Fuel Tank Sheath Instal	VS-40320
Main Instrument Board Instal	VS-40411
Man Bomb & Drop Tank Release	VS-44693
Oxygen System Instal	VS-40550
Pedal, Rudder & Brake Controls	VS-40410
Pilot's Seat Instal	VS-40448
Radio Control Box Instal	VS-40538
Relief Tube Instal	VS-44025
Shell Assem L H	VS-40231
Shell Assem R H	VS-40232
Signal Pistol Support Instal	VS-40556
Sliding Section Control Instal	VS-40248
Sliding Section Instal	VS-40211
Stick & Rudder Pedal Lock Assem	VS-44500
Stick & Torque Tube Controls	VS-40405
Tail Wheel Lock Ind Controls	VS-40417-2
Windshield Defrosting Instal	VS-40441-1
Wing Flap Controls	VS-45004

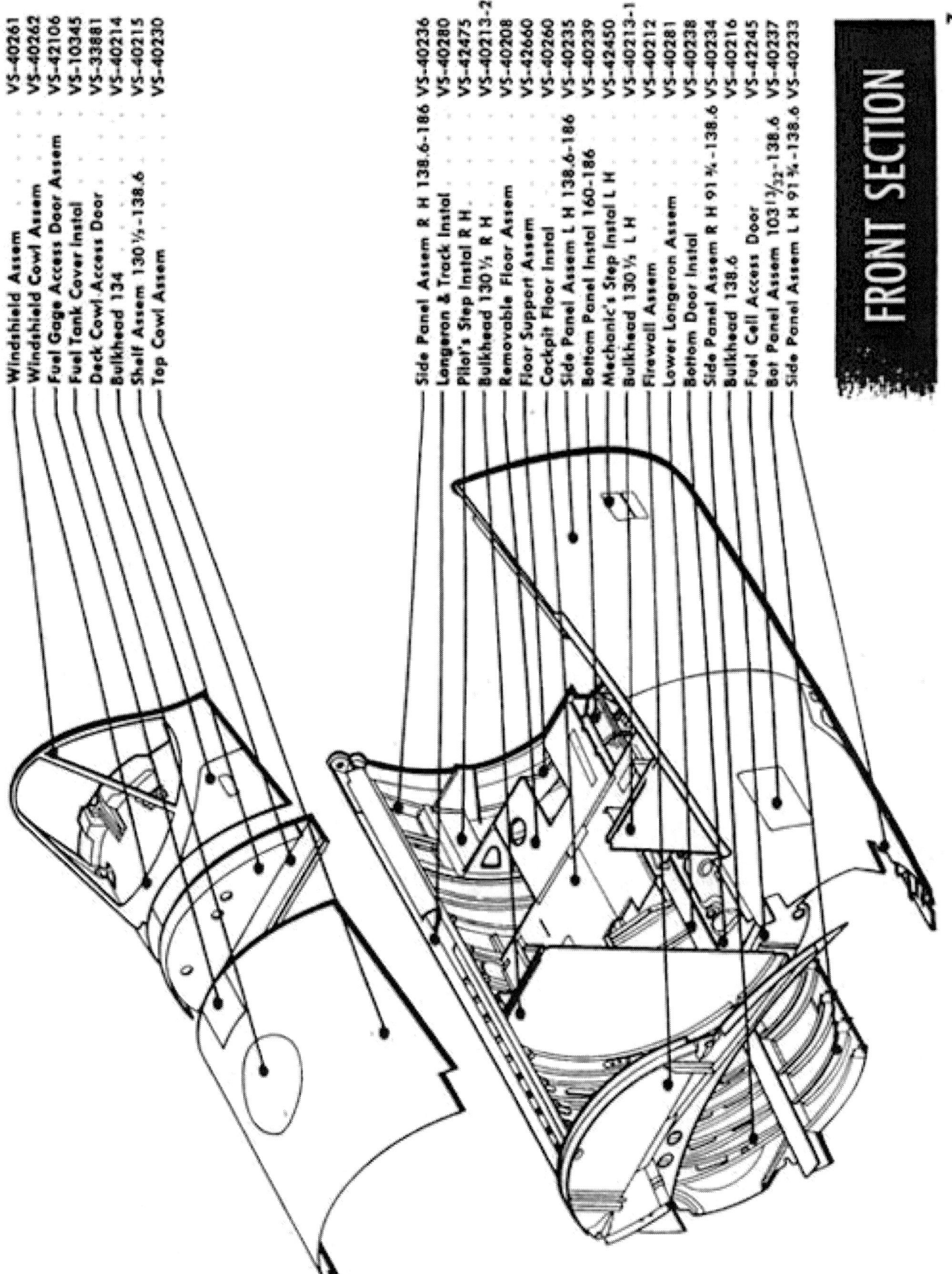

Windshield Assem	VS-40261
Windshield Cowl Assem	VS-40262
Fuel Gage Access Door Assem	VS-42106
Fuel Tank Cover Instal	VS-10345
Deck Cowl Access Door	VS-33881
Bulkhead 134	VS-40214
Shelf Assem 130½-138.6	VS-40215
Top Cowl Assem	VS-40230
Side Panel Assem R H 138.6-186	VS-40236
Longeron & Track Instal	VS-40280
Pilot's Step Instal R.H.	VS-42475
Bulkhead 130½ R H	VS-40213-2
Removable Floor Assem	VS-40208
Floor Support Assem	VS-42660
Cockpit Floor Instal	VS-40260
Side Panel Assem L H 138.6-186	VS-40235
Bottom Panel Instal 160-186	VS-40239
Mechanic's Step Instal L H	VS-42450
Bulkhead 130½ L H	VS-40213-1
Firewall Assem	VS-40212
Lower Longeron Assem	VS-40281
Bottom Door Instal	VS-40238
Side Panel Assem R H 91¾-138.6	VS-40234
Bulkhead 138.6	VS-40216
Fuel Cell Access Door	VS-42245
Bot Panel Assem 103 13/32-138.6	VS-40237
Side Panel Assem L H 91¾-138.6	VS-40233

FRONT SECTION

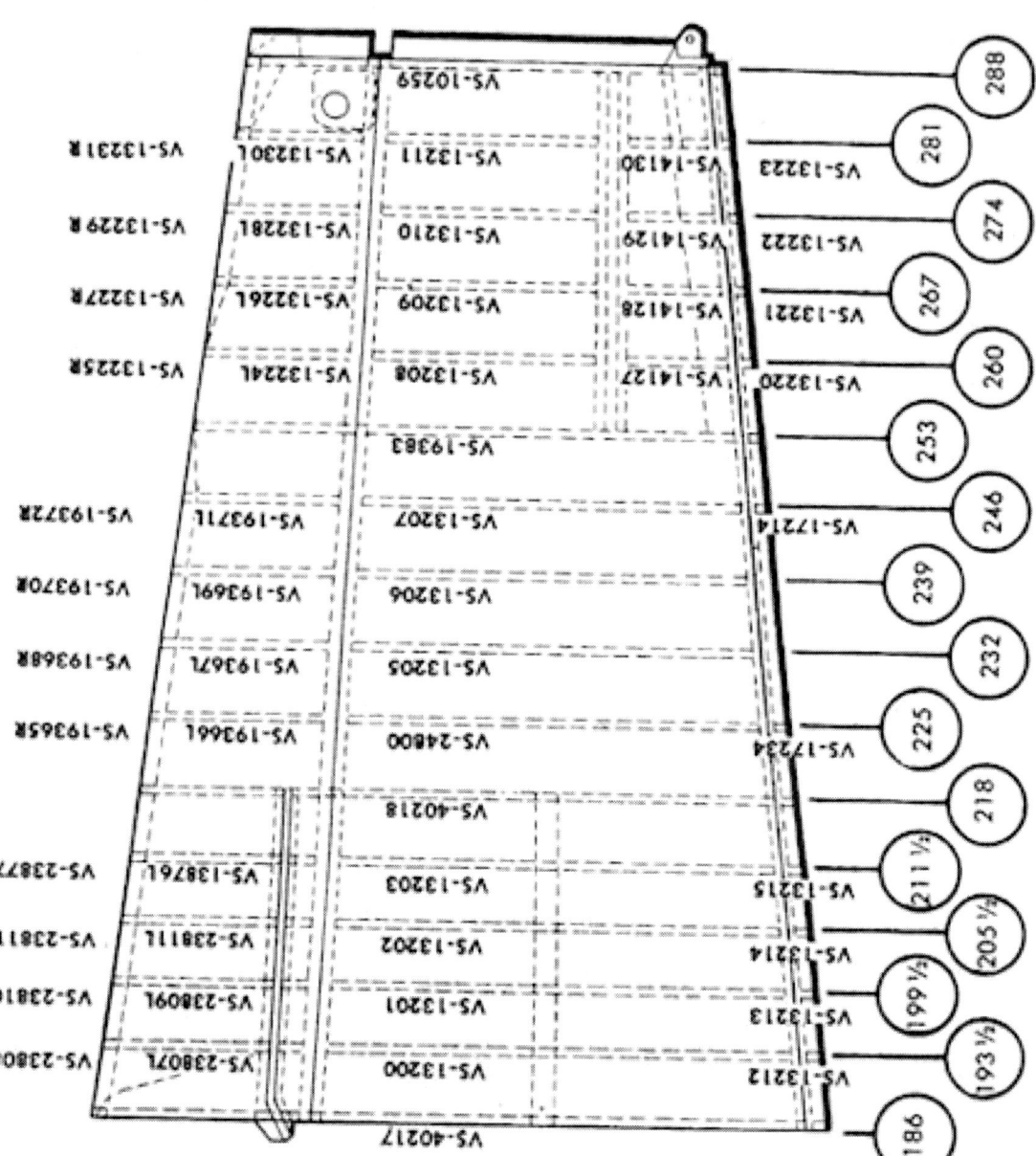

MID SECTION

MID SECTION

Part	Number
Mid Section Fuse Assem	VS-40228
Fumeproofing—Mid Section	VS-40294
Fumeproofing—Blkhd 218	VS-40279
Bulkhead 288	VS-10259
Aft Turtledeck Assem 218-288	VS-19360-1
Lift Tube Instal	VS-10818
Top Longeron 259½-288	VS-12254-3
Top Longeron 186-253	VS-19359
Upper Longeron Assem L H	VS-12244-2
Upper Longeron Assem R H	VS-12244-1
Side Panel Assem R H	VS-40242
Head Rest Instal	VS-40496
Turtledeck Panel Assem 186-218	VS-23804
Bulkhead 253	VS-19383
Bottom Panel Assem—246-288	VS-11834-7
Track Instal	VS-11203
Baggage Compartment Instal	VS-45013
Bulkhead 218	VS-40218
Radio Compartment Floor	VS-33800-2
Bottom Access Door	VS-17243
Lower Longeron Assem R H	VS-12245-1
Lower Longeron Assem L H	VS-12245-2
Bottom Panel Assem 186-225	VS-40267
Bulkhead 186	VS-40217
Side Panel Assem L H	VS-40241
Armor Plate Instal	VS-48760
Electrical System Instal	VS-40432
Fumeproofing Sta 218	VS-40279
Hydraulic System Instal	VS-40276
Reel Instal	VS-44876
Tow Target Release Instal	VS-44022
Turtledeck Assem Complete	VS-40245

AFT SECTION

Aft Section Fuselage VS-40229
Rear Fuselage Assem VS-40209
Rear Fuselage Unit VS-40210
Tail Cone Assem VS-19504

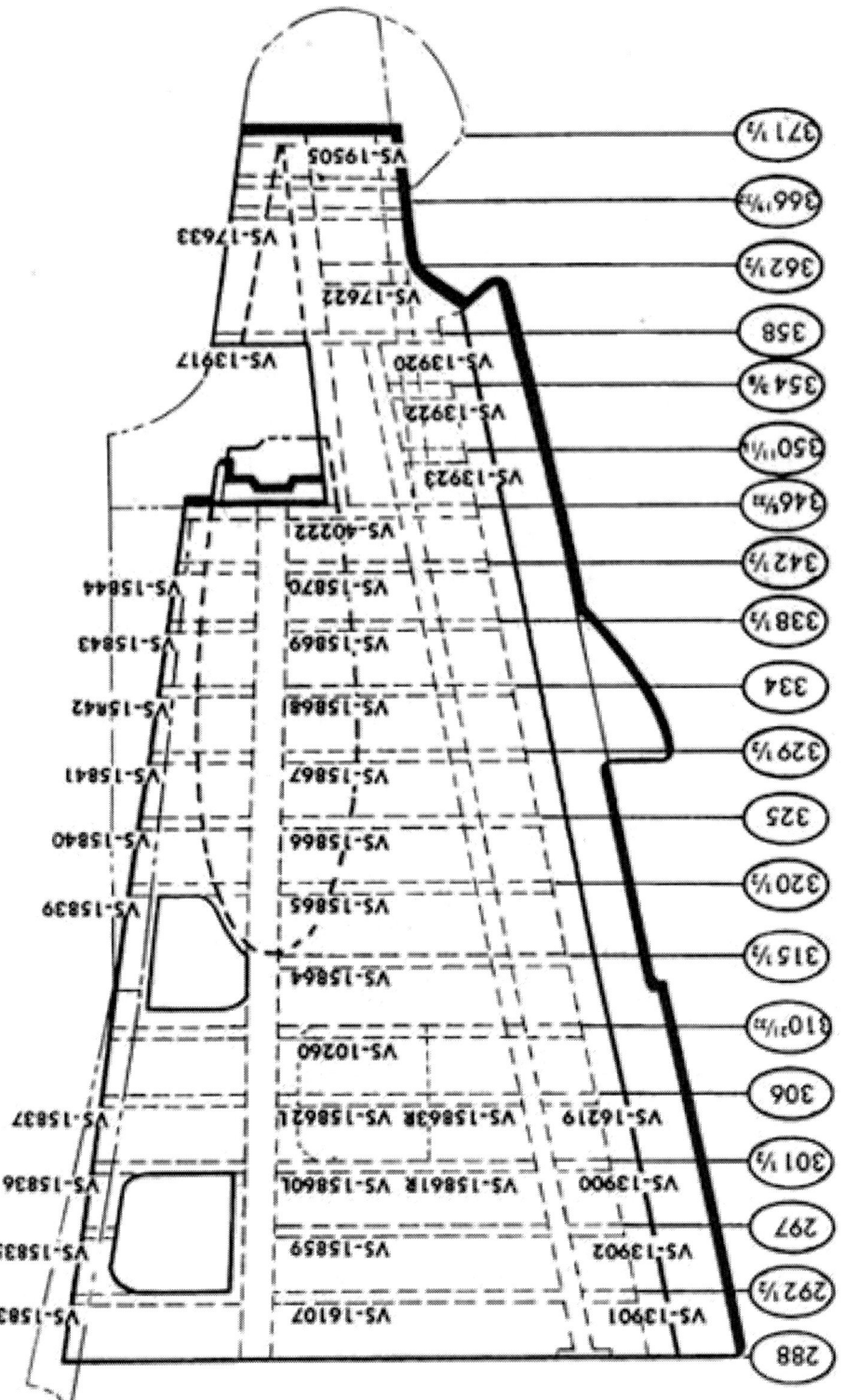

AFT SECTION

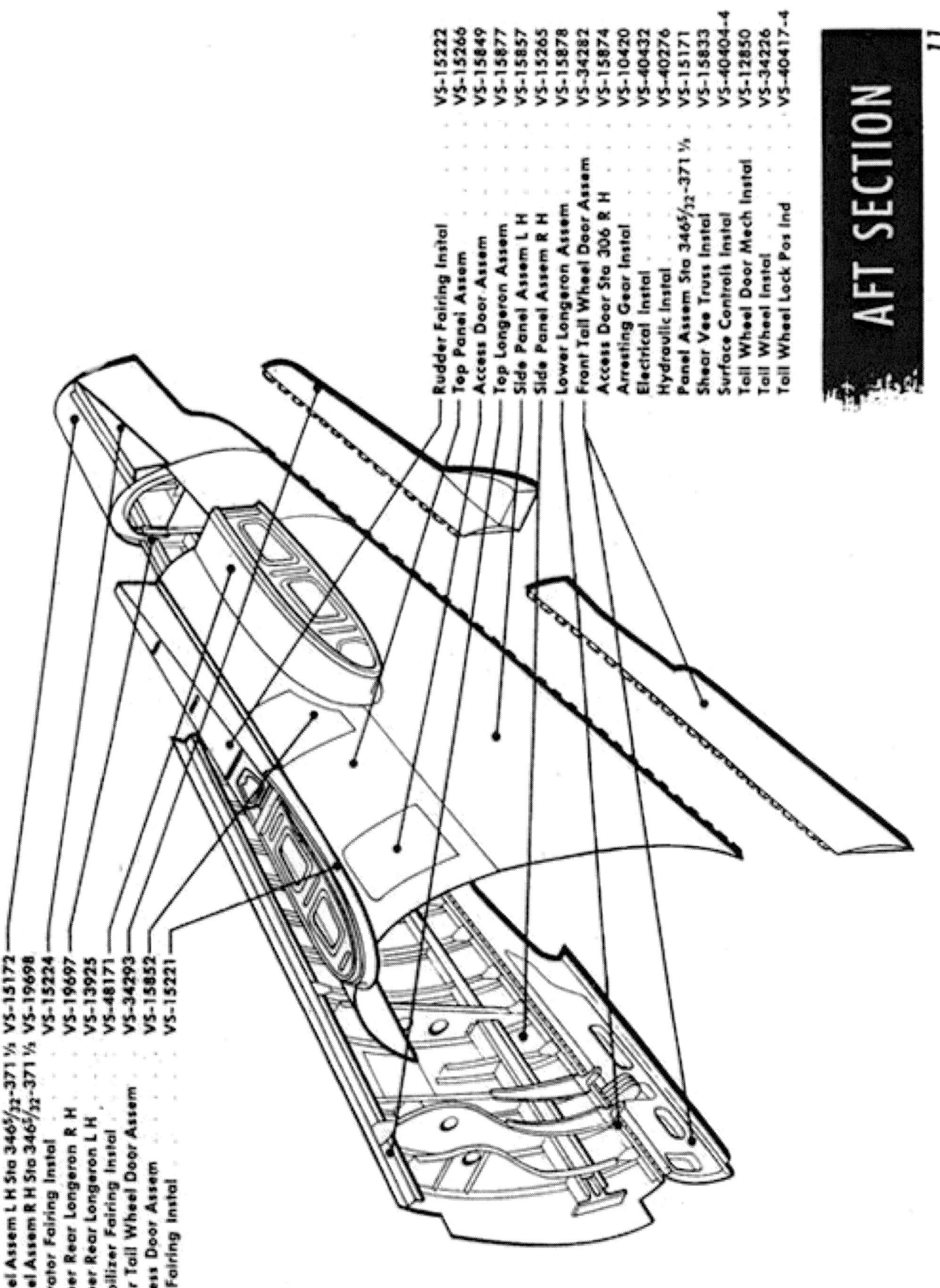

Panel Assem L H Sta 346⁵/₃₂-371 ½	VS-15172
Panel Assem R H Sta 346⁵/₃₂-371 ½	VS-19698
Elevator Fairing Instal	VS-15224
Upper Rear Longeron R H	VS-19697
Upper Rear Longeron L H	VS-13925
Stabilizer Fairing Instal	VS-48171
Rear Tail Wheel Door Assem	VS-34293
Access Door Assem	VS-15852
Fin Fairing Instal	VS-15221
Rudder Fairing Instal	VS-15222
Top Panel Assem	VS-15266
Access Door Assem	VS-15849
Top Longeron Assem	VS-15877
Side Panel Assem L H	VS-15857
Side Panel Assem R H	VS-15265
Lower Longeron Assem	VS-15878
Front Tail Wheel Door Assem	VS-34282
Access Door Sta 306 R H	VS-15874
Arresting Gear Instal	VS-10420
Electrical Instal	VS-40432
Hydraulic Instal	VS-40276
Panel Assem Sta 346⁵/₃₂-371 ½	VS-15171
Shear Vee Truss Instal	VS-15833
Surface Controls Instal	VS-40404-4
Tail Wheel Door Mech Instal	VS-12850
Tail Wheel Instal	VS-34226
Tail Wheel Lock Pos Ind	VS-40417-4

CENTER SECTION

Center Section Assem	VS-40770
Aileron Controls Instal	VS-10475
Electrical Instal	VS-40429
Fuel System Instal	VS-40325-3
Hinge Pin Access Door	VS-24086
Hydraulic System Instal	VS-40272
Interbeam Assem	VS-40776
Landing Gear Doors Instal	VS-13593
Leading Edge & Beam Assem	VS-40775
Lubricating System Instal	VS-40307-3
Manual Bomb & Tank Release	VS-44693
Wing Folding Gap Door Instal	VS-10794
Lower Airduct Panel Assem	VS-40764
Pylon Instal	VS-34025
Catapult Hook Instal	VS-12795
Oil Cooler Door Assem	VS-40757
Shoulder Instal	VS-14419
Lower Inboard Skin Assem	VS-40733
Movable Fairing Instal	VS-10857
Inboard Fixed Fairing	VS-14994
Inb'd Landing Gear Door Assem	VS-10781
Outboard Fixed Fairing	VS-14995
Outb'd Landing Gear Door Assem	VS-10775
Lower Outb'd Skin Assem & Instal	VS-24091
Flap Gap Closing Door Instal	VS-14655

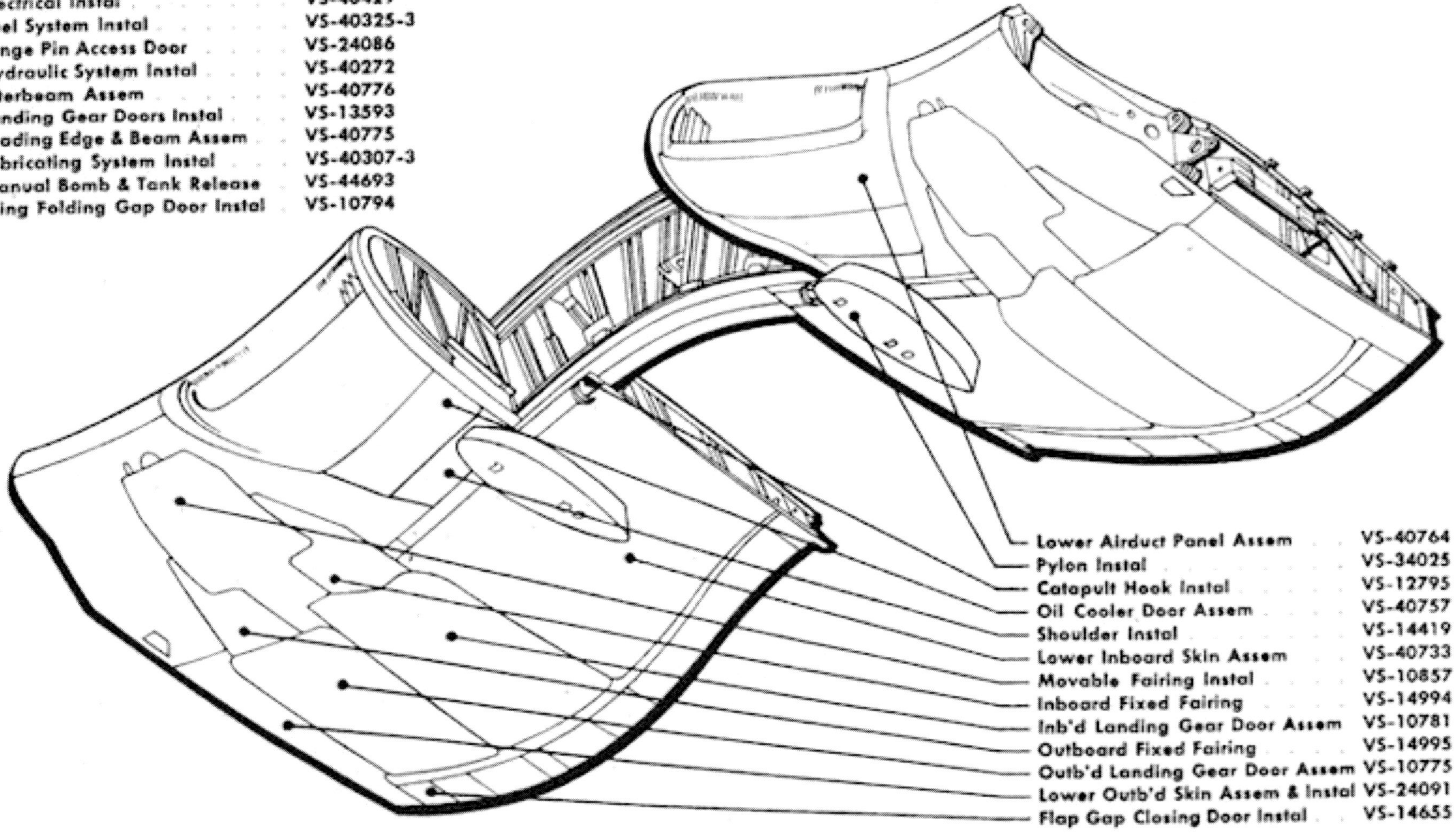

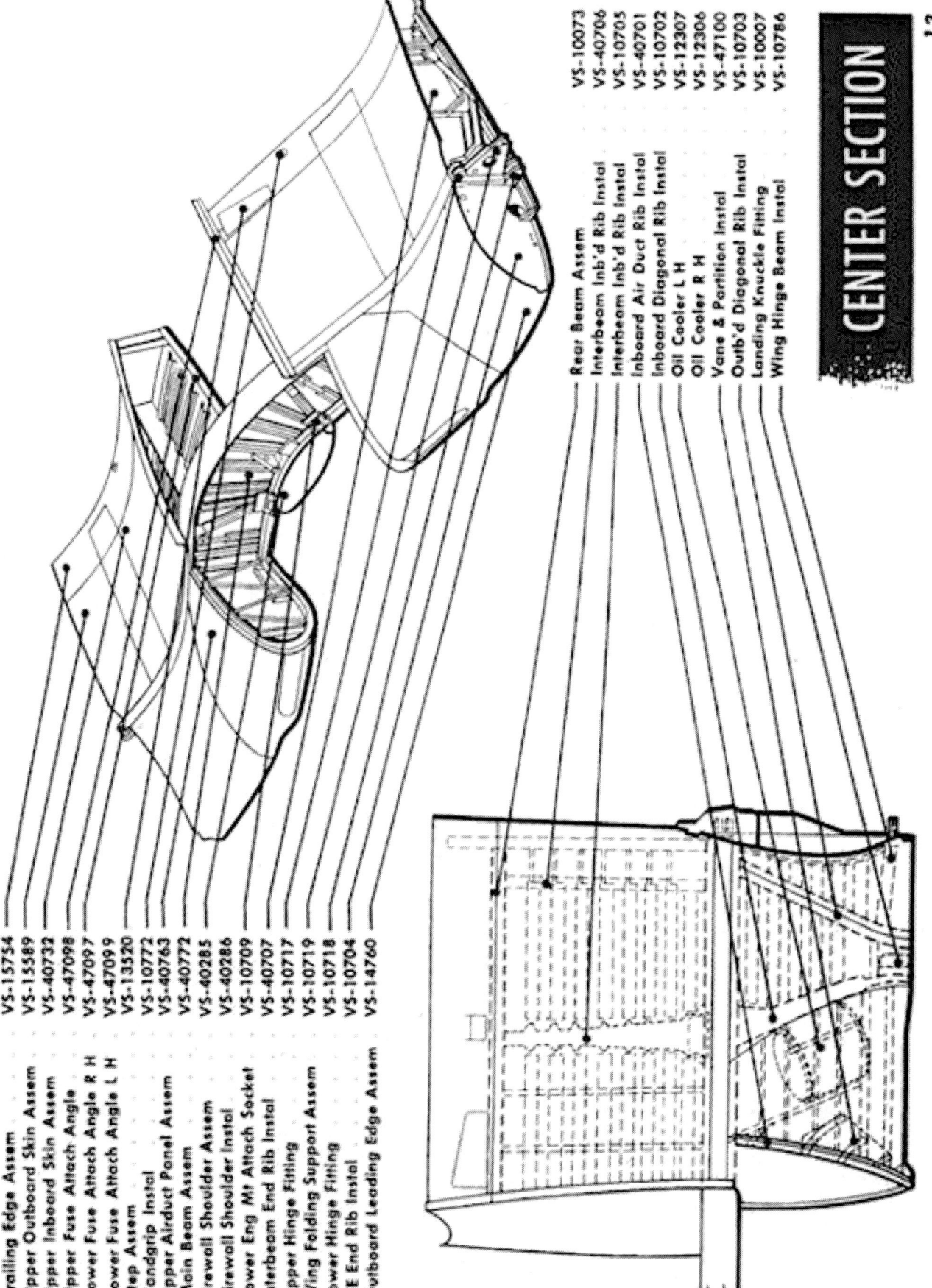

CENTER SECTION

Trailing Edge Assem	VS-15754
Upper Outboard Skin Assem	VS-15589
Upper Inboard Skin Assem	VS-40732
Upper Fuse Attach Angle	VS-47098
Lower Fuse Attach Angle R H	VS-47097
Lower Fuse Attach Angle L H	VS-47099
Step Assem	VS-13520
Handgrip Instl	VS-10772
Upper Airduct Panel Assem	VS-40763
Main Beam Assem	VS-40772
Firewall Shoulder Assem	VS-40285
Firewall Shoulder Instl	VS-40286
Lower Eng Mt Attach Socket	VS-10709
Interbeam End Rib Instl	VS-40707
Upper Hinge Fitting	VS-10717
Wing Folding Support Assem	VS-10719
Lower Hinge Fitting	VS-10718
L E End Rib Instl	VS-10704
Outboard Leading Edge Assem	VS-14760
Rear Beam Assem	VS-10073
Interbeam Inb'd Rib Instal	VS-40706
Interbeam Inb'd Rib Instal	VS-10705
Inboard Air Duct Rib Instal	VS-40701
Inboard Diagonal Rib Instal	VS-10702
Oil Cooler L H	VS-12307
Oil Cooler R H	VS-12306
Vane & Partition Instal	VS-47100
Outb'd Diagonal Rib Instal	VS-10703
Landing Knuckle Fitting	VS-100007
Wing Hinge Beam Instal	VS-10786

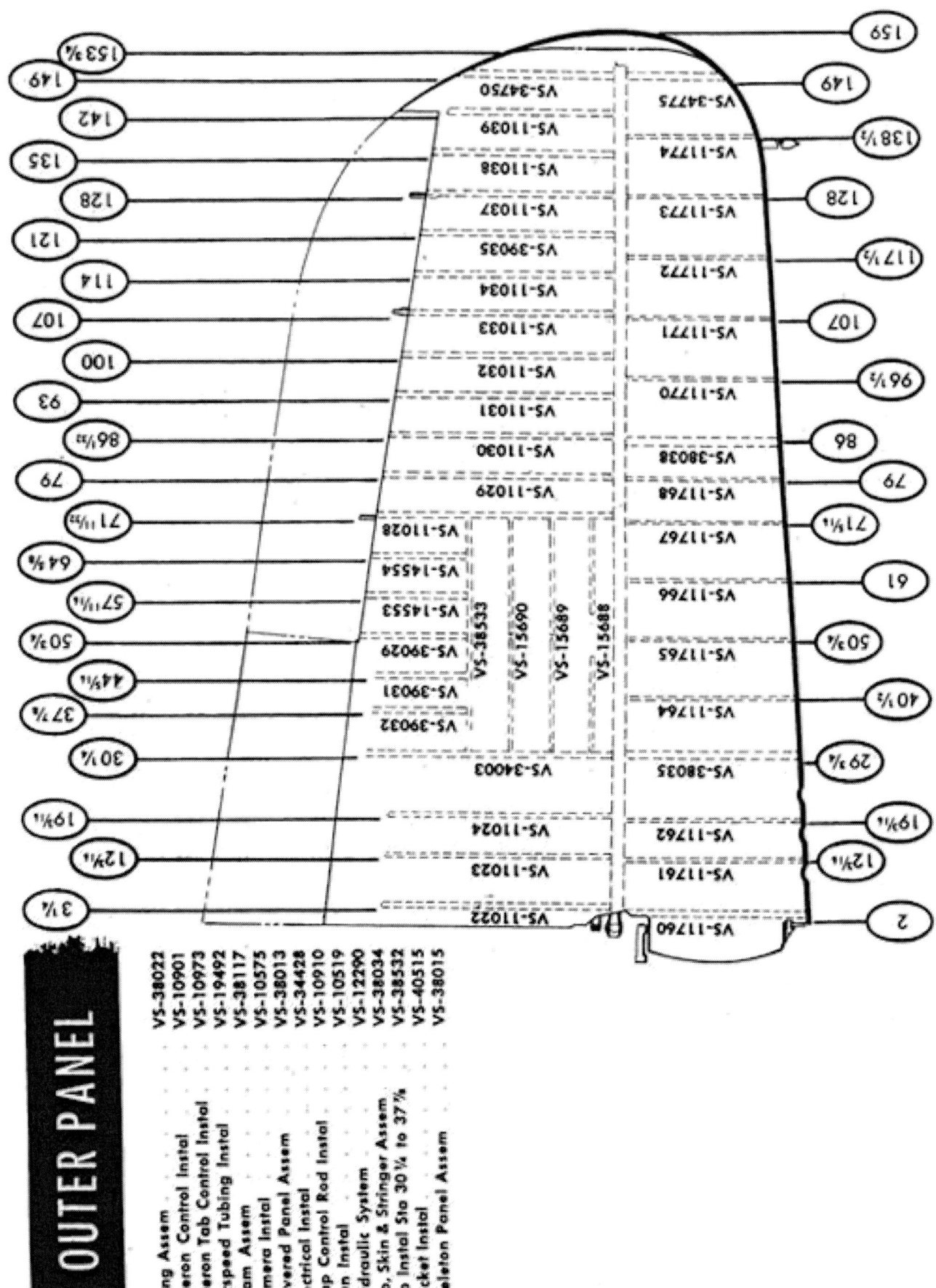

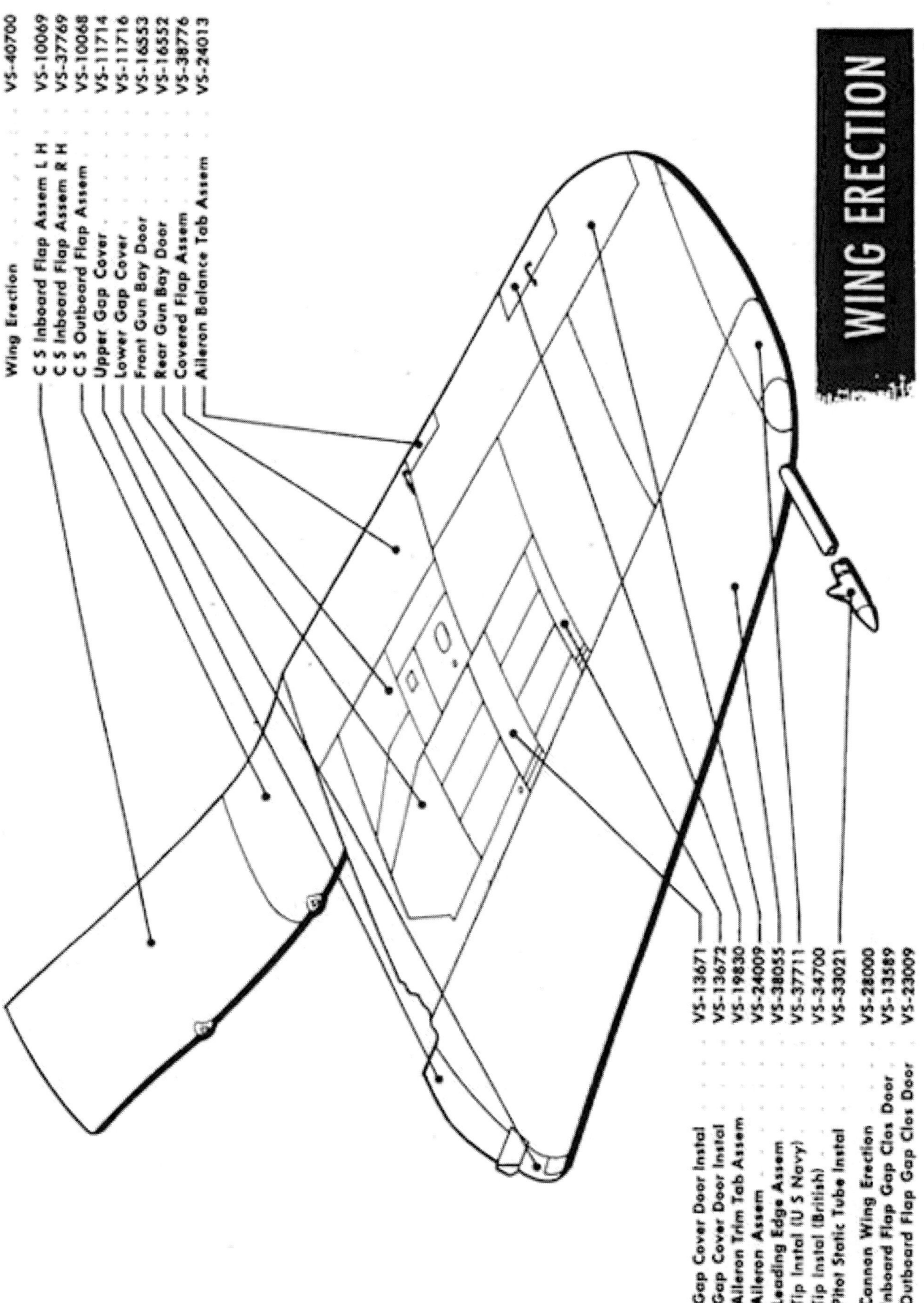

WING ERECTION

Wing Erection	VS-40700
C S Inboard Flap Assem L H	VS-10069
C S Inboard Flap Assem R H	VS-37769
C S Outboard Flap Assem	VS-10068
Upper Gap Cover	VS-11714
Lower Gap Cover	VS-11716
Front Gun Bay Door	VS-16553
Rear Gun Bay Door	VS-16552
Covered Flap Assem	VS-38776
Aileron Balance Tab Assem	VS-24013
Gap Cover Door Instal	VS-13671
Gap Cover Door Instal	VS-13672
Aileron Trim Tab Assem	VS-19830
Aileron Assem	VS-24009
Leading Edge Assem	VS-38055
Tip Instal (U S Navy)	VS-37711
Tip Instal (British)	VS-34700
Pitot Static Tube Instal	VS-33021
Cannon Wing Erection	VS-28000
Inboard Flap Gap Clos Door	VS-13589
Outboard Flap Gap Clos Door	VS-23009

15

TAIL ERECTION

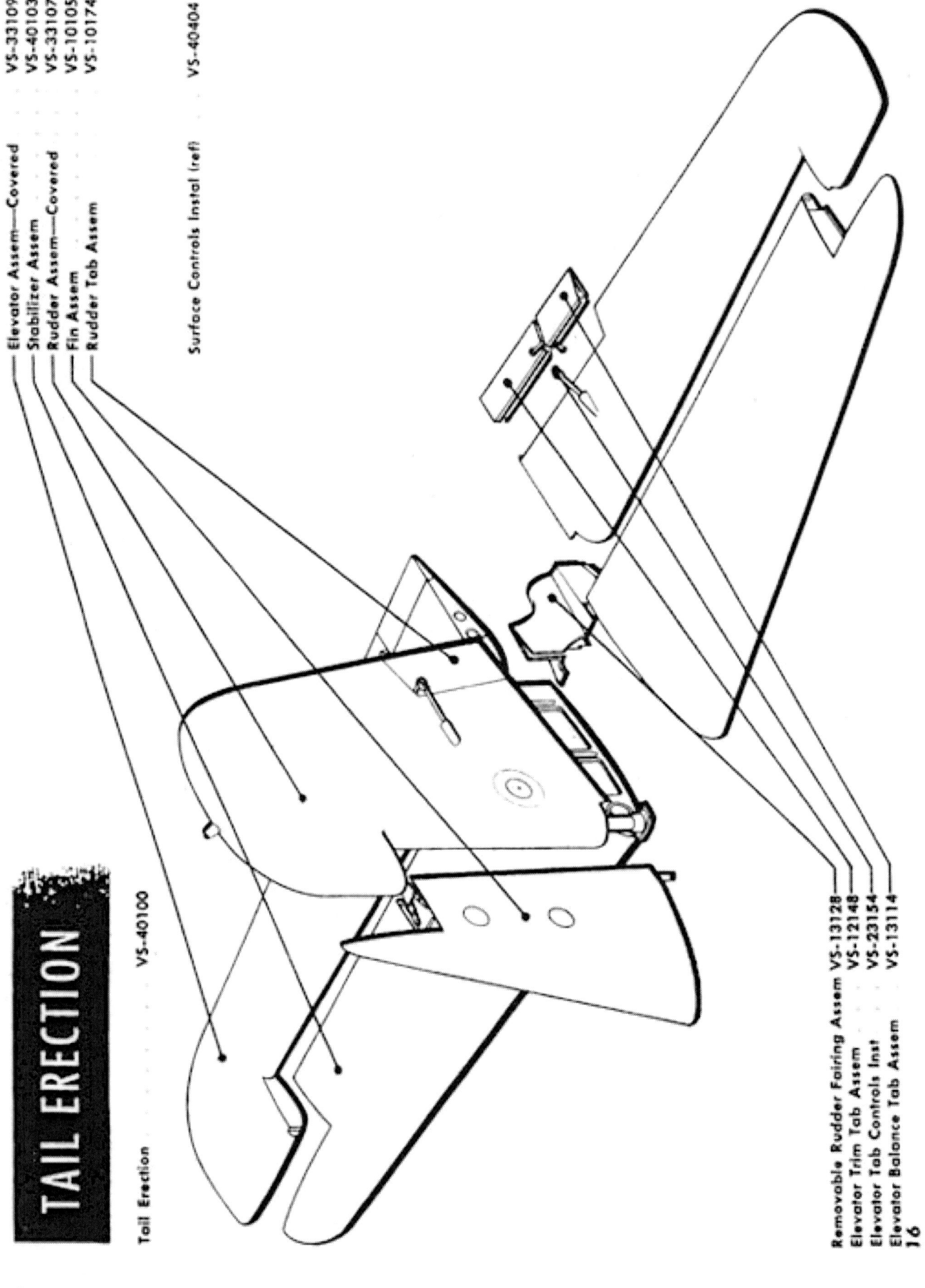

Tail Erection VS-40100

Elevator Assem—Covered	VS-33109
Stabilizer Assem	VS-40103
Rudder Assem—Covered	VS-33107
Fin Assem	VS-10105
Rudder Tab Assem	VS-10174

Surface Controls Instal (ref) VS-40404

Removable Rudder Fairing Assem	VS-13128
Elevator Trim Tab Assem	VS-12148
Elevator Tab Controls Inst	VS-23154
Elevator Balance Tab Assem	VS-13114

16

THE FLEET TYPE SUBMARINE
The definitive technical guide -- profiling the submarines of World War II
RESTRICTED
NOW ALSO AVAILABLE FROM
WWW.PERISCOPEFILM.COM:
THE DEFINITIVE GUIDE TO THE
SUBMARINES THAT WON WWII
Originally published as NAVPERS 16160
June 1946
WWW.PERISCOPEFILM.COM

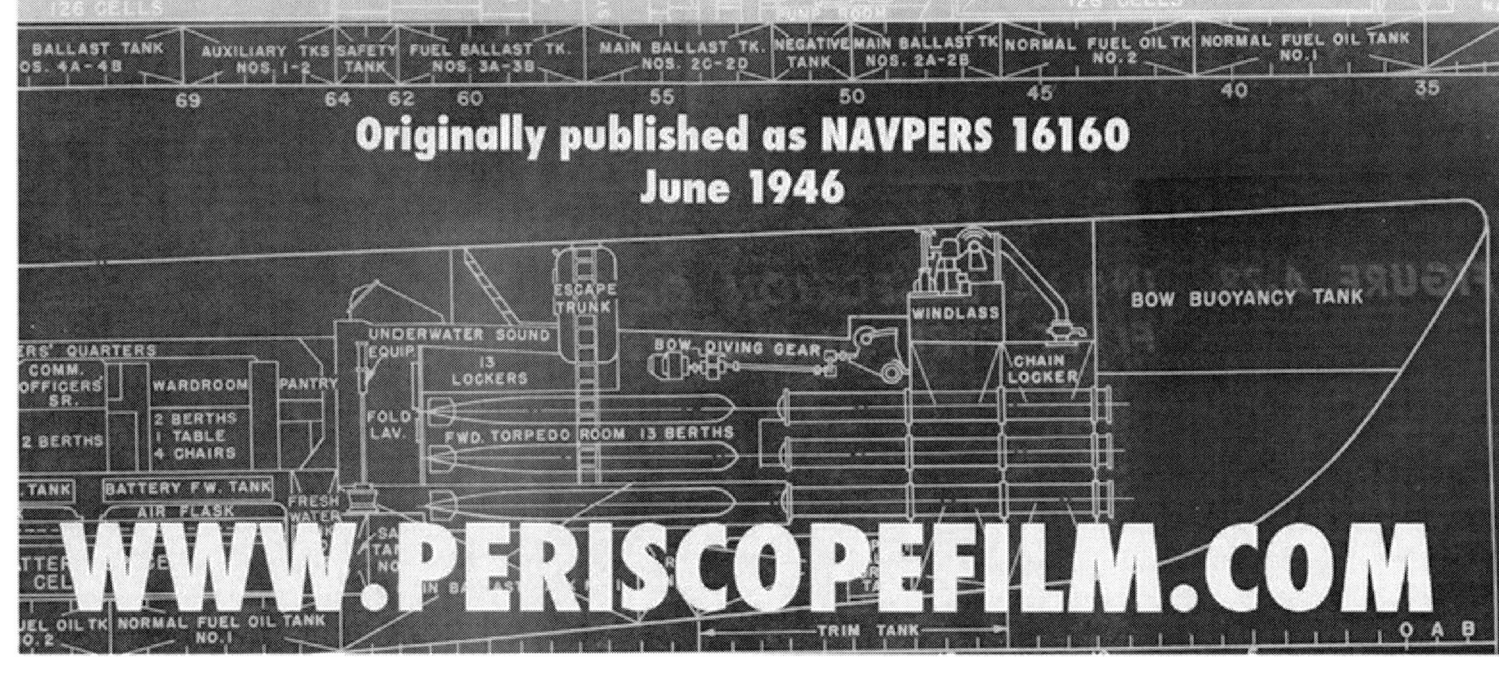

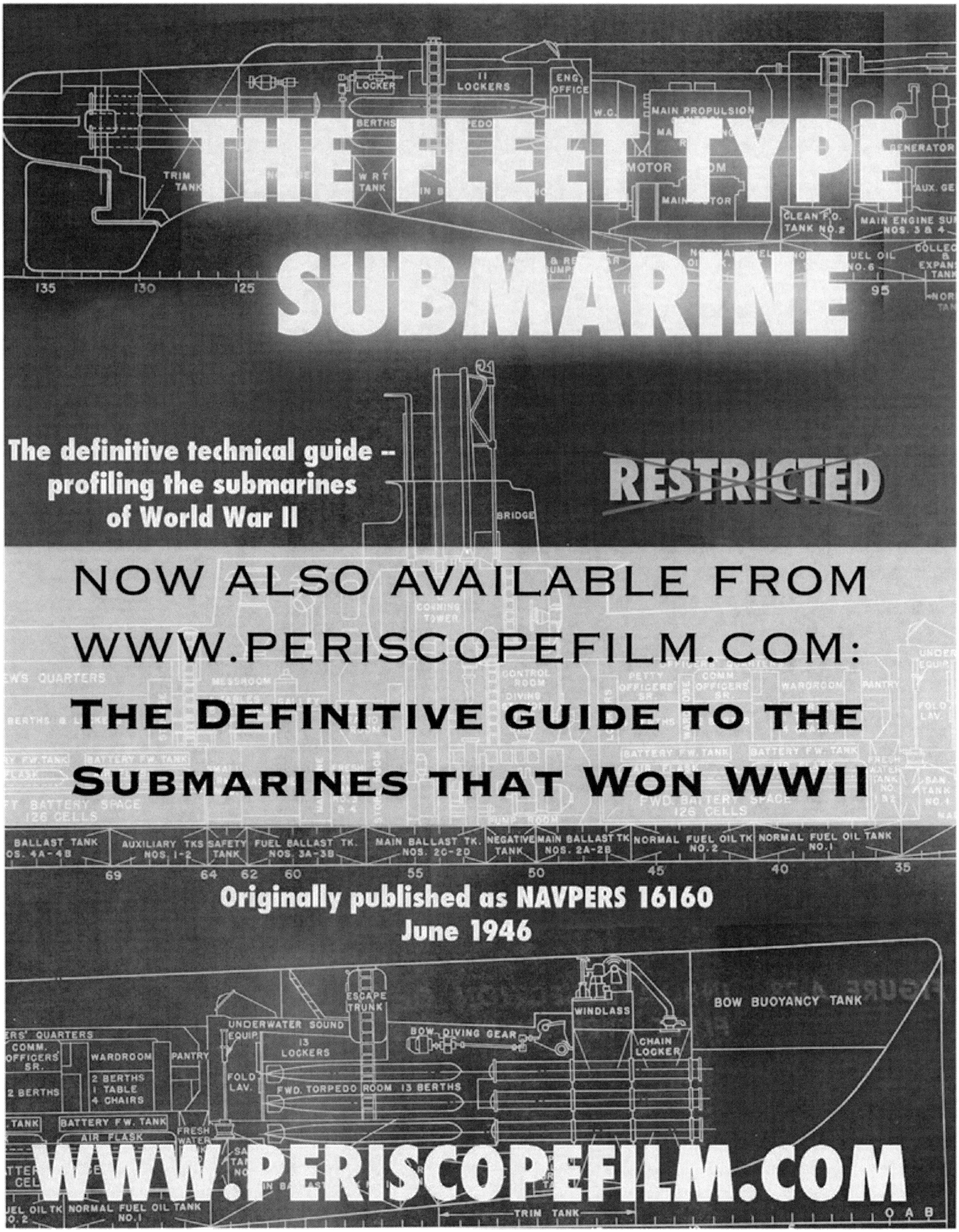

HISTORIC SUBMARINE, DEEP SEA DIVING AND RESEARCH DVDs NOW AVAILABLE FROM WWW.PERISCOPEFILM.COM

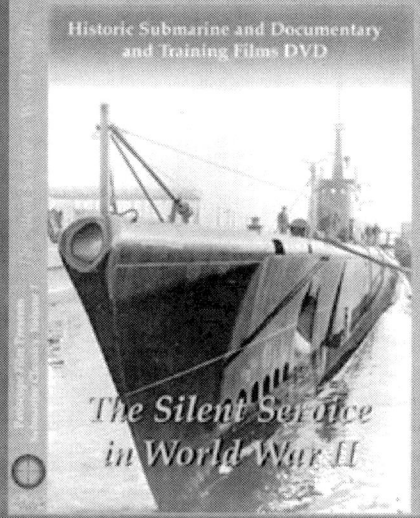

FEATURING HISTORIC FILMS RESCUED FROM U.S. NAVY ARCHIVES!

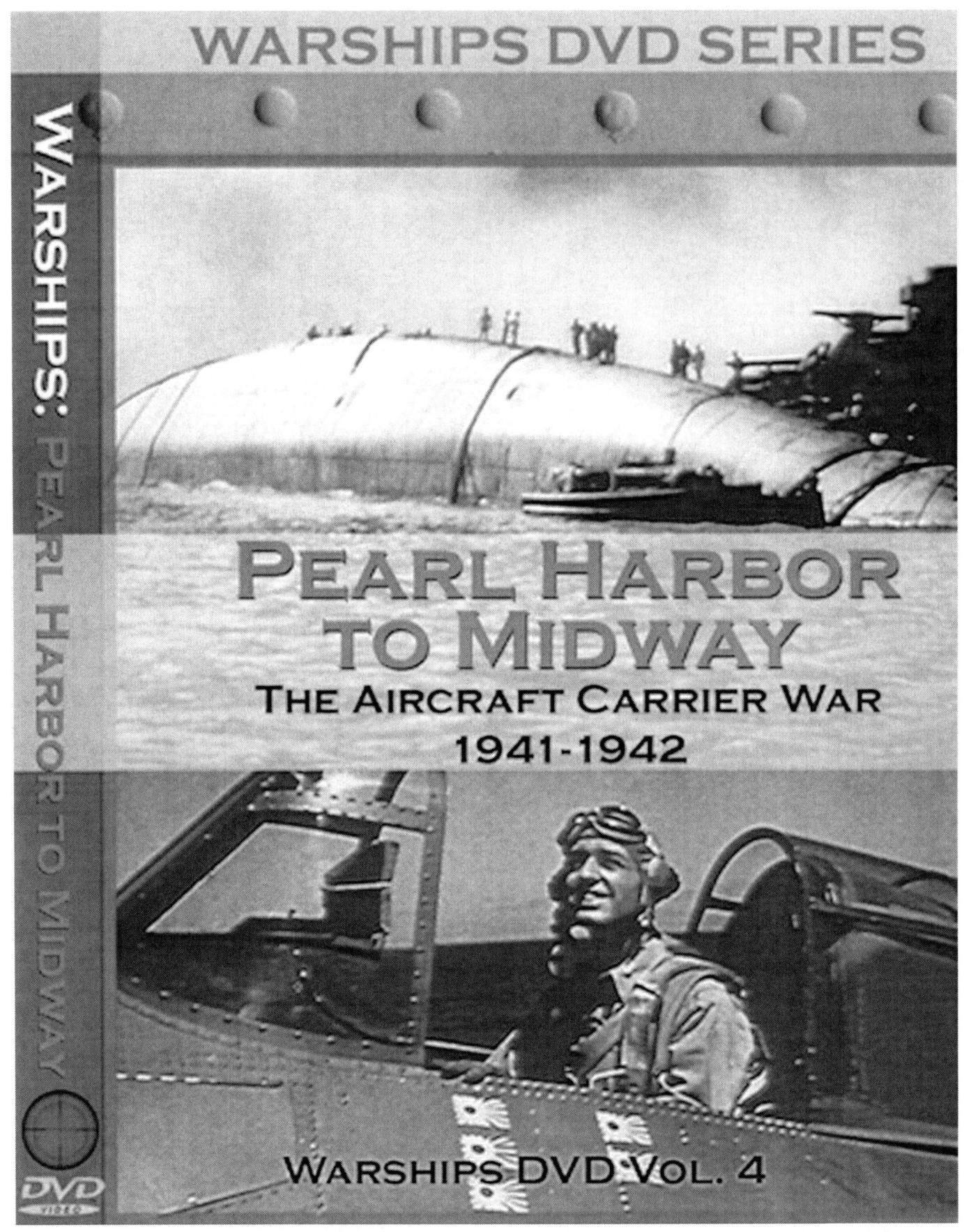

NOW AVAILABLE!

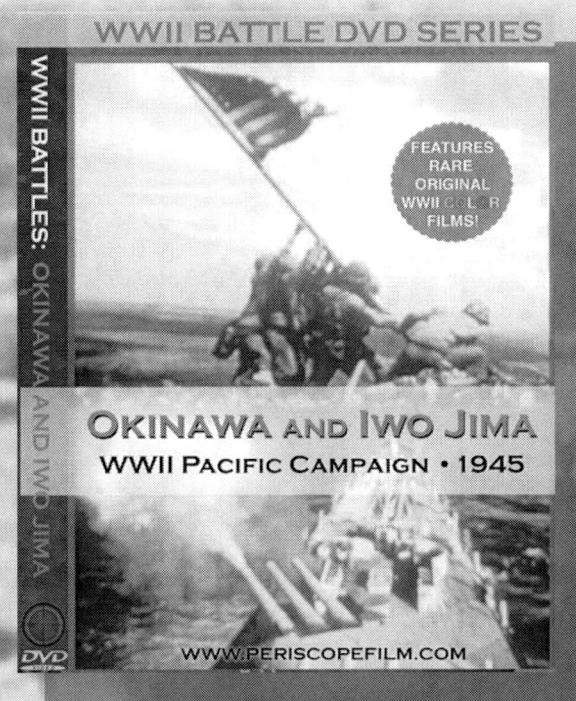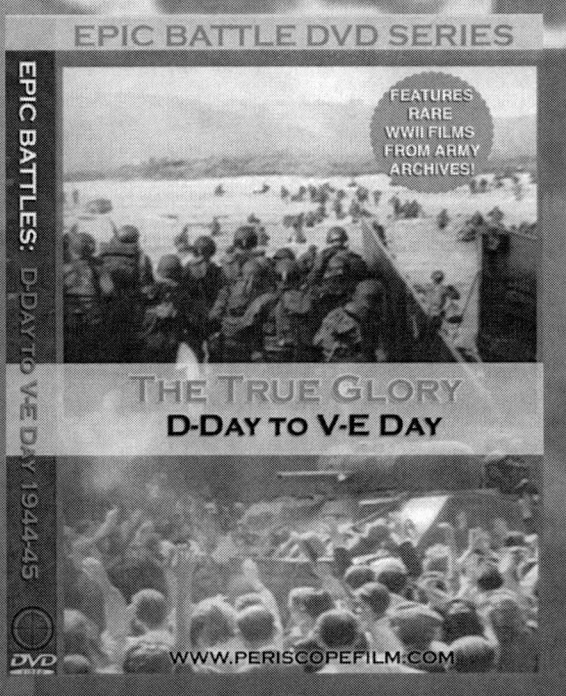

Warships DVD Series

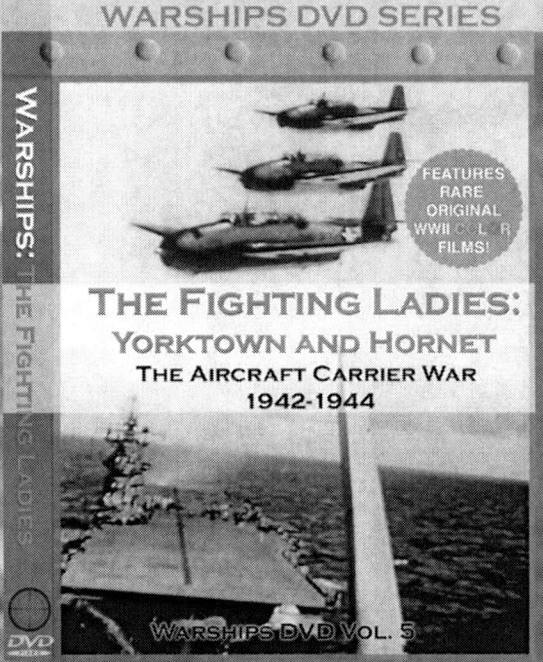

Now Available!

Aircraft At War DVD Series

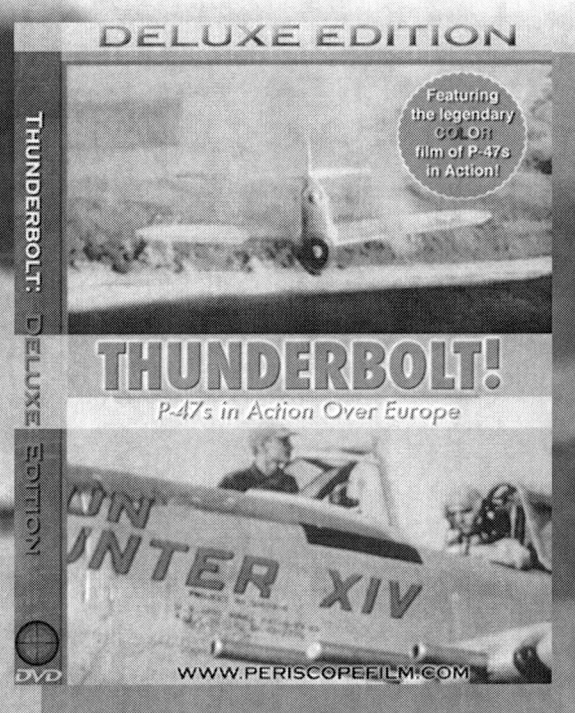

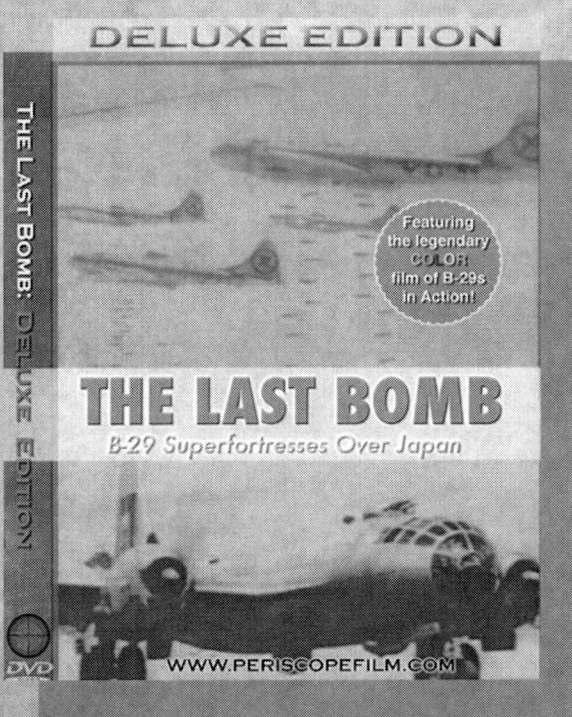

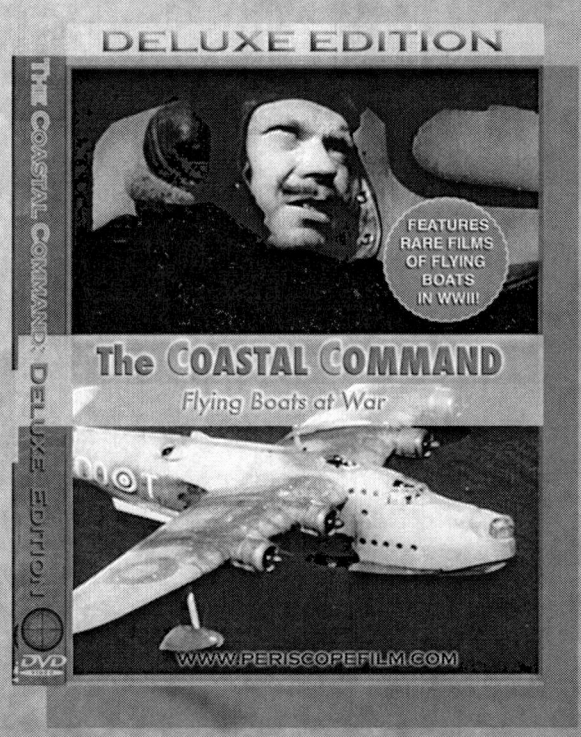

Now Available!

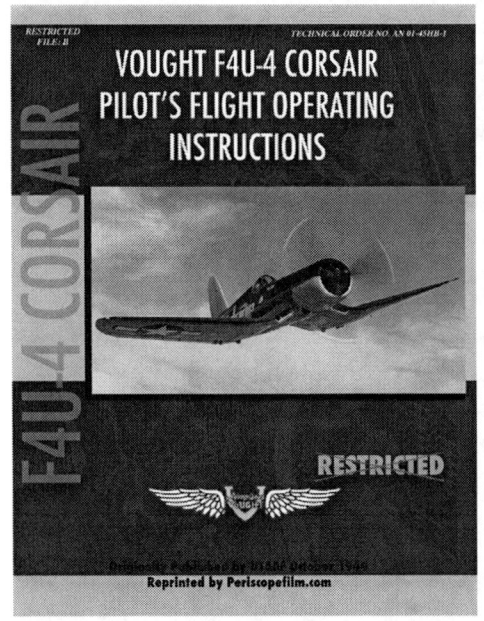
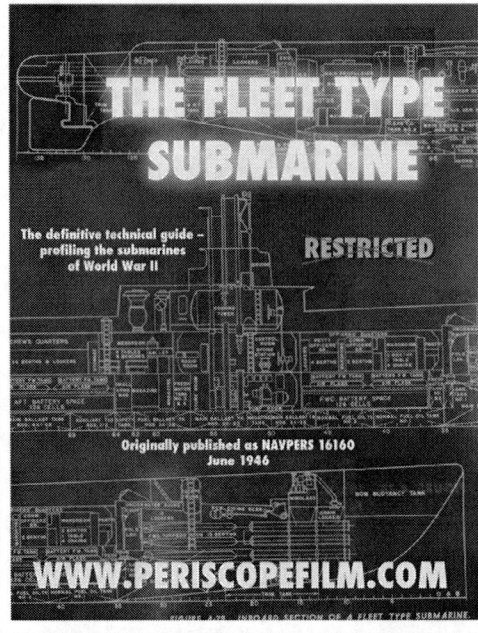
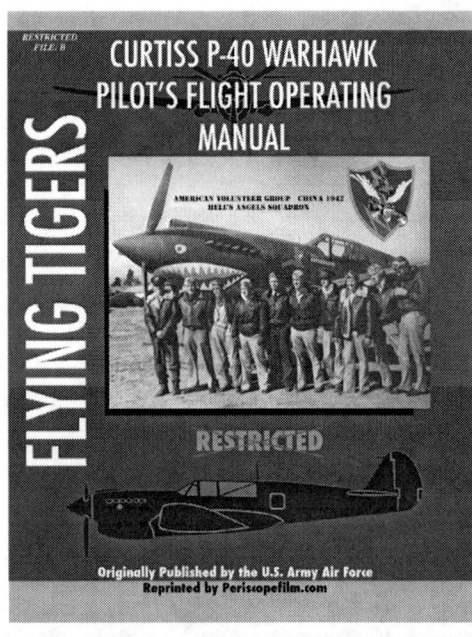
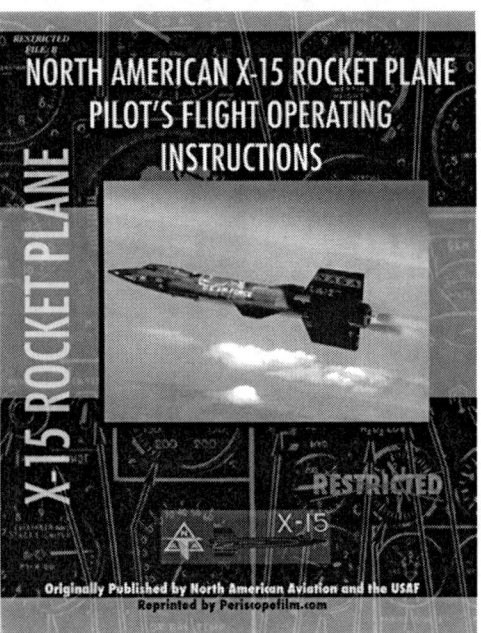
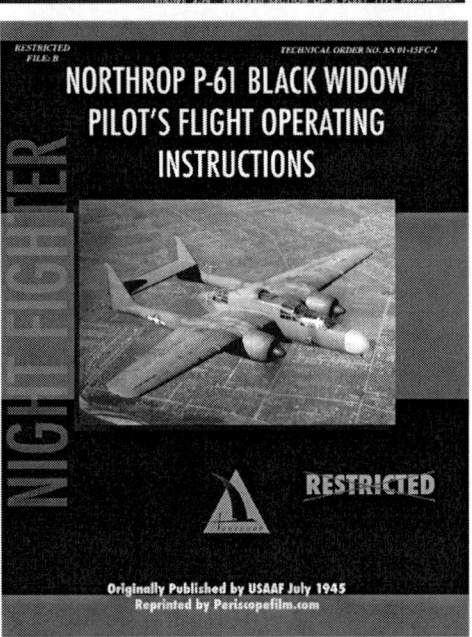
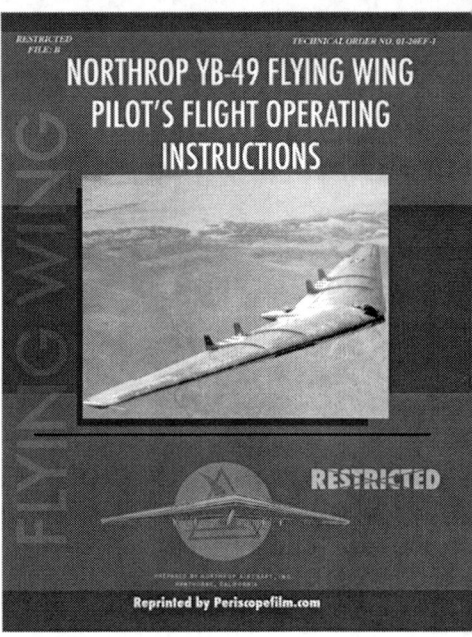

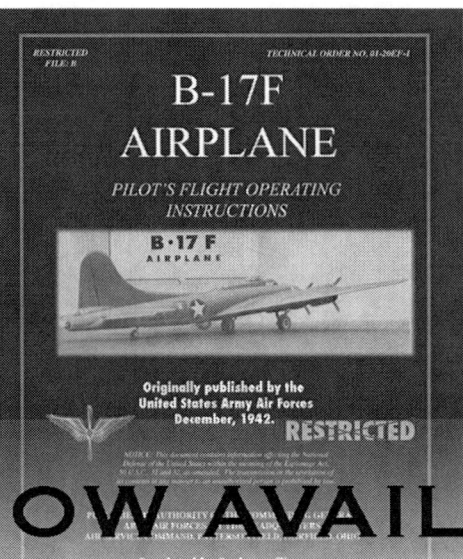

ALSO NOW AVAILABLE FROM PERISCOPEFILM.COM

Vought F4U-4 Corsair
Pilot's Flight
Operating Instructions
©2006, 2009 Periscope Film LLC
All Rights Reserved

www.PeriscopeFilm.com
ISBN 978-1-4116-8960-2

CPSIA information can be obtained at www.ICGtesting.com
Printed in the USA
LVOW10s0049171014

409058LV00002B/61/P